ORIGIN of UNIVERSE

ORIGIN *of* UNIVERSE

THE LIGHT OF SYSTEM PHILOSOPHY

SCIENCE AND PHILOSOPHY

GEORGE LUKE

Copyright © 2018 by George Luke.

ISBN: Softcover   978-1-5437-0416-7
              eBook   978-1-5437-0417-4

All rights reserved. No part of this book may be used or reproduced by any means, graphic, electronic, or mechanical, including photocopying, recording, taping or by any information storage retrieval system without the written permission of the author except in the case of brief quotations embodied in critical articles and reviews.

Because of the dynamic nature of the Internet, any web addresses or links contained in this book may have changed since publication and may no longer be valid. The views expressed in this work are solely those of the author and do not necessarily reflect the views of the publisher, and the publisher hereby disclaims any responsibility for them.

First published by PGL BOOKS as eBook in Amazon Kindle Store

Print information available on the last page.

To order additional copies of this book, contact
Partridge India
000 800 10062 62
orders.india@partridgepublishing.com

www.partridgepublishing.com/india

Also by George Luke

SAPTALOKADARSHANAM SAMGRAHAM (MALAYALAM)
JEEVANUM PARINAMAVUM – SYSTEM PHILOSOPHIYUDE VELICHAM (MALAYALAM)
LIFE AND MIND: THE LIGHT OF SYSTEM PHILOSOPHY
DISCOVERY OF REALITY: THE LIGHT OF SYSTEM PHILOSOPHY

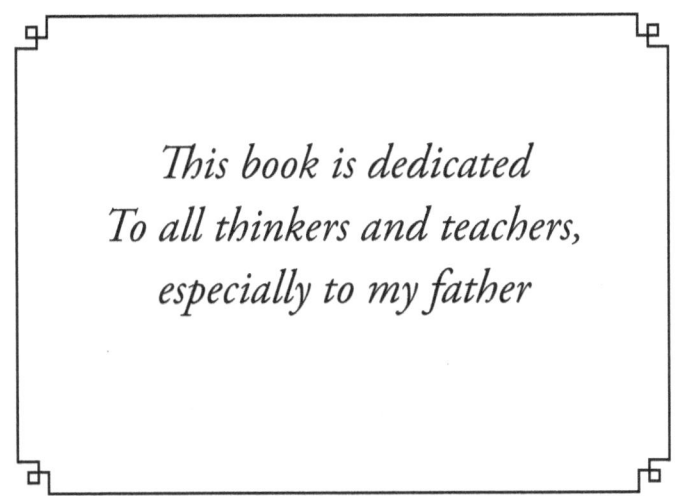

*This book is dedicated
To all thinkers and teachers,
especially to my father*

CONTENTS

Preface xi

Prologue xix

Introduction xxvii

Chapter 1: Fundamental Ideas of Physical Science 1

   1.1 Main Areas of Science 2

   1.2 Classical Physical Science 8

   1.3 Quantum Mechanics - Key Concepts. 10

   1.4 Quantum Field Theory 20

   1.5 Doctrines of Cosmology 27

Chapter 2: Big Bang Cosmology 30

   2.1 Standard Big Bang Theory 32

   2.2 Inflationary Big Bang Theory 37

Chapter 3: Quantum Cosmology 41

   3.1 Quantum Cosmology-3. (Level III) 46

   3.2 Quantum Cosmology-2 (Level II) 46

      3.2.1 Quantum Gravity (the period up to 10^{-35} seconds ABB) 49

      3.2.2 Inflation (10^{-35} seconds after big bang) 53

3.2.3 Symmetry Breaking and Higgs Mechanism 55

3.3 Quantum Cosmology -1 (Level I) 58

    3.3.1 Membrane Theory 58

    3.3.2 Multiverse Theory 59

    3.3.3 Dark Matter and Dark Energy 61

3.4 Critical Summary and Outstanding Issues 67

Chapter 4: Introducing Philosophy of Science 76

4.1 What is Scientific Method? 79

4.2 Epistemology of Classical Science 86

    4.2.1 Methodology and Source : 86

    4.2.2 Justification and Truth in Classical Science 92

4.3 Epistemology of Quantum Mechanics and Big Bang Cosmology 92

    4.3.1 Logical Positivism 93

    4.3.2 Justification and Truth of Quantum Mechanics 97

4.4 Philosophy of Quantum Cosmology -- Special Issues 97

    4.4.1 Methodology of Quantum Cosmology 97

    4.4.2 Source : Materialist Philosophy of Mind 104

4.5 The Crisis in Philosophy of Physical Science 105

Chapter 5: System Philosophy of Science 110

5.1 Summary of Dilemmas in Methodology and Source 111

5.2 System Philosophy about Methodology and Source 115

5.3 Overview of Problem of Justification in Physical Science ... 119

5.3.1 Classical Science 119

5.3.2 Quantum Mechanics...122

5.3.3 Quantum Cosmology.. 125

5.4 System Model of Justification - Existence of
Physical World 128

5.5 Spiritual Science of Ancient and Medieval Periods............ 135

Chapter 6: The Cosmological Puzzles Finally Solved 148

6.1 Space, Time and Gravitational Waves................................. 149

6.2 Solution to the Dilemma about Matter............................. 161

6.3 Frequently Asked Question about the Origin of Universe.. 170

6.4 Do We Need Intelligent Design Argument? 176

Chapter 7: Further Vision of System Philosophy 183

7.1 Summary of the Book : *Life and Mind*...............................184

- Life and Evolution..184

- System Philosophy of Mind...187

- System Philosophy of God and Evil.............................189

- Science- Religion Synthesis..190

7.2 Summary of the Book : *Discovery of Reality*......................192

  o A Guide to the Levels of Knowledge 192

  o Philosophy and its Main Divisions 193

  o World and Reality 195

  o Existence of Seven Life Systems................................. 197

  o Comprehensive View about Truth 199

o   Glimpse of the Path Ahead..201

Bibliography... 205

Index of Names... 213

Index of Subjects 217

Preface

From our scientific as well as common sense view, the concepts like matter, space and time would stand for the reality of physical world. Yet scientists and thinkers have been struggling to understand what these entities actually are. In modern physics, the fundamental components of matter are defined on the basis of mathematical models. Besides, space and time are measurable variables. Can we say that these aspects of physical reality have existence in a similar way as visible things exist? What do we mean by the statement that space and time started with the event of Big Bang?

For explaining the origin of universe in a scientific way, we have to essentially deal with the beginning of matter, space and time; this is the field of *cosmology*, in which the universe is conceived in physical sense. Accordingly, this book is primarily aimed to address the profound questions of cosmology. In other words, it is a journey for analyzing the nature of scientific investigation and inference pertaining to the origin of physical world. We seek the merger of philosophy and science – this subject is specifically called *philosophy of science* - resulting in the philosophical treatment of scientific knowledge.

As mentioned above, science deals with the universe expressed in *physical terms* involving matter and energy. However, in recent decades, many interesting publications have emerged for taking a critical view about the enterprise of science in spite of its wonderful discoveries. In his book *The Trouble with Physics* (2008), Lee Smolin describes the failure of quantum physics in answering the fundamental questions about the existence of physical world. The origin of matter, or rather its constituents, seems to be beyond the field of experimental methods.

Can we believe in the speculations using mathematical models such as strings, membranes and multiverse, which add extra dimensions to space and time? What kind of truth is there in scientific narratives about the origin of universe? There are still more outstanding problems "that remain unanswered because of the incompleteness of the twentieth century's scientific revolution".

Rupert Sheldrake, in *The Science Delusion* (2013), tries to expose the misleading ideas in the theoretical foundation of science. According to him, in a summary way, the drawbacks of scientific theory are twofold. First is the assertion that the real stuff of universe is matter. Hence scientific method is built upon the philosophy of materialism. Consequently, the aspects of life and consciousness are treated as byproduct or epiphenomenon of material process in the body of living beings. This amounts to the denial of non-physical aspects like purpose and creativity in nature. Obviously, it is the opposite of idealism, which holds that Mind exists as a real substance.

Then, the second drawback of scientific theory is the exclusion of nonscientific (mental, psychic or paranormal) phenomena from its purview. The occurrence of telepathy and precognition, for example, are despised just like religious faith as well as traditional practices of medicine. Concerned with the mental aspect of universe, Sheldrake leans on the notion of *self-organizing systems*, in tune with the empirical and mystical view of a section of contemporary scientists.

Reading and reflecting upon such critical books, I enquired how to fit the mystical theory of systems within the framework of philosophical thought. Throughout my self-study of philosophy, spanning three decades, I desired passionately to develop a treatise for solving the basic issues of science and other kinds of knowledge. It is gratifying to state that my quest is fructified in the form of an innovative vision named as System Philosophy. Brief record of my intellectual development is separately given under the forthcoming *Prologue*.

System Philosophy is to be distinguished from the *systems philosophy* or *systems view*, popularized in the writings of a group of thinkers including mainly David Bohm, Fritjof Capra and Ervin Laszlo. They have followed the empirical approach adopting the process view for articulating that the universe consists of *self-organizing systems,* which are wholes of interrelated components; these systems have been organized at various levels primarily due to the particle-wave (matter-energy) duality of subatomic phenomena. Obviously the *systems philosophy* is a description of the activity of reality, without implying existence.

In contrast, *system philosophy* is defined here as an integrative thought about the universe as a system of matter and consciousness, where these constituents are in dialectical and productive relation. It effectively shows that things exist by the union of opposites. This philosophical perspective about real existence involves the synthesis of rational and empirical aspects of knowledge. I can simply and tentatively state that the conflict between the principal visions of Reality runs through the history of philosophy. Can we synthesize these divergent streams? This is the central question of my series of three books -- *Origin of Universe, Life and Mind* and *Discovery of Reality.*

With this background, the present book formulates the System Philosophy of Science, in which existence is represented by the mathematical technique of X-Y coordinates. These innovative perspectives can synthesize the rational and empirical aspects of knowledge.

The unique features of this book can be picked up by comparing the text with the relevant sections of the books used for reference as included in bibliography. **All chapters have a considerable set of original ideas; these are clearly marked by the symbol [*]. The main examples of innovative ideas are listed below chapter wise.**

Chapter 1 : *Fundamental Ideas of Physical Science.* The special feature of this chapter is the focus on the distinctive paradigms or worldviews of science for describing *philosophy of science.* The clear definition of *Physical Science,* pertaining to modern period, would place it in contrast to the ancient and medieval kinds of knowledge about nature. Mentioning about the controversies in quantum mechanics, I have given innovatively the true meaning of particle-wave duality as following: *Every particle has inherent energy.* This meaning will help us to tackle the ontological problems of physical world.

Chapter 2 : *Big Bang Cosmology.* This chapter describes the theoretical issues of big bang theory and introduces the term *refined big bang theory* as an original idea. The information from the book of Paul Davies is used to work out the Table titled Expansion and Diameter of Universe.

Chapter 3 : *Quantum Cosmology.* This chapter gives many new ideas; important ones are given below.

- Two visible and three invisible levels of physical world
- Three levels of Quantum Cosmology, which are further divided into six stages.
- The author suggests as following: the exclusion of graviton from standard model is now compensated by accepting Higgs boson as the fundamental exchange particle causing the gravitational force. The confusion about graviton can be safely dispensed with and, at the same time, the lacuna of standard model is removed.
- The comprehensive list of drawbacks of Quantum Cosmology

Chapter 4 : *Introducing Philosophy of Science.* Scientific method has five stages called Theory, Hypothesis, Deduction, Testing and Induction, which are abbreviated as Ty, H, D, T and I respectively. The phrase *TyHDTI scheme* is used to denote the scientific method; it will be seen later that it is the common method for all kinds of knowledge. The three stages of Ty, H and D constitute rational knowledge, while the stages of T and I constitute empirical knowledge. The distinction between content view and process view has important application throughout the book. We have to analyze all disciplines in accordance with the 2x2 Table of content view, process view, rational view and empirical view. This point will be elaborated in the separate book *Discovery of Reality.*

We know that classical science adopts content view; its theories are based on the principles of mechanistic worldview. In the case of quantum mechanics, the subatomic phenomena have the wonderful property of particle-wave duality, necessitating the physical process view. This chapter analyses the methodological issues of the two paradigms of physical science. In a ground-breaking manner, I propose that the methodology of *quantum cosmology* is Modern Phenomenalism (simply, Mophism) following the agenda of materialism and empiricism. *The methodological dilemma in this stage is similar to that of classical science.* Further, the terms like *a priori*, underdetermination, modern phenomenalism and scientism are explained along with analysis of the main doctrines and conflicts in philosophy of science.

Chapter 5 : *System Philosophy of Science.* In this ground-breaking chapter, recognizing the incomplete state of philosophy of science, I would propose the principles of System Philosophy of Science in order to derive the innovative aspects of scientific *methodology, source and justification.* The most important points are: (1) the philosophical definition of system, (2) efficacy of a good theory in reducing the problem of induction, (3) criticism on scientific realism and (4) the synthesis of matter and energy through X-Y model to show the phenomenal existence of physical world. *These ideas would remove the agnosticism of Immanuel Kant and give the justification of scientific laws.* Additionally, the notion of past universe, the depiction of big bang event and the modification of balloon model of expanding universe to show the space outside our present universe are also original ideas.

Further it is explained that scientists resort to the application of Aristotle's *rules of thought* for talking about the existence of individual entities like matter, energy, electron and quark. In the final section, we would clear the ambiguities about the nature of ancient knowledge like *Ayurveda* and *Yoga* by using the paradoxical phrase *spiritual science*, which is my original idea. The curious issue of paranormal phenomena also is resolved using the system model human mind.

Chapter 6 : *Cosmological Puzzles Finally Solved.* Specific examples of originality are the following:

> ➤ The controversies about the notions of space and time are explained in a unique manner. By linking *arrow of time* with

cause-effect, we can note that time always move from left to right. Accordingly, the notion of *negative time* is introduced logically. Further, the presence of *dark matter* and *dark energy* also must be taken into account for explaining the curvature of space-time.
- Einstein's thought experiments like twin paradox, time dilation, space-time curvature and gravitational force, time warp and time travel are criticized noting that the great scientist resorted to naïve realism. On the basis of System Philosophy of Science, we can propose four notions of space and time – *visible, abstract, quantum* and *astrophysical*. These concepts are based on the **layered perspective** of physical world.
- The key principle is that the physical world exists like a factory in the form of matter-energy system; it solves the conundrum of God and Intelligent Designer. But physical world has phenomenal existence only.
- Finally the FAQ about the origin of universe is answered without resorting to the Intelligent Designer Argument.

Chapter 7 : *Further vision of System Philosophy*. Having solved the longstanding issues of physical science, this chapter gives a glimpse of the topics elaborated in the two companion books titled respectively as *Life and Mind* and *Discovery of Reality*.

This book is written in a pedagogic style, by explaining all concepts in a basic manner making use of suitable Tables and Diagrams. So the book would serve as text as well as reference for all persons, having at least undergraduate level of education, who desire to know the answers for philosophical questions about basic aspects of any discipline.

Specifically, the System Philosophy about quantum physics, life, evolution, mind, reality and social systems are the original features of this book. The two central principles advanced here are:

- ■ *Everything is a system of opposite forces having material and mental aspects respectively.*

- *We are not entitled to assume realism - metaphysical or scientific – in our knowledge about reality and phenomena.*

As per this path, I have tried to rediscover the theoretical framework of philosophy in order to make it efficient for solving the problems of our life.

Prologue

MY EXPERIMENTS WITH PHILOSOPHY

My life story has many unexpected twists and turns, which reveals the specific intentions of *almighty, the reality of universe*. This brief autobiographical note is intended to reveal the trials and tribulations, which I experienced in the course of my philosophical quest. I was born on 03 June 1953 as the eldest son of Aleykutty and P. L. George belonging to Puthankulam family of Thodupuzha Taluk in Kerala state, south India. Father was a teacher of Malayalam language. I am a member of Roman Catholic Church of Christianity.

Starting life in an underdeveloped part of Kerala state, I studied in Malayalam medium schools - at Neyyassery (1958–62) up to fourth standard and afterwards at Kodikulam (1962-68). In the primary and middle levels I was very poor in mathematics. But during the year of eighth standard I had a sudden interest in Geometry, particularly in proving theorems. It kindled my aptitude for mathematics, which in coming years helped to expand my brain power. I passed SSLC in 1968 meritoriously, having placed at fiftieth rank in the all-Kerala list. I received the national merit scholarship also.

During the vacation after SSLC examination, I started reading books available in local library. Having good proficiency in reading and writing essays, I curiously read books on popular science including the articles that appeared in periodicals like Mathrubhumi weekly, especially relating to Quantum Physics and Cosmology. This interest in scientific topics prompted me to think on the deeper aspects of world and to brood over philosophical questions. My mathematical and logical mind became suitable for creative thinking on the ultimate issues, though I did not have formal education in philosophy.

I graduated from Newman College (1968-73) at Thodupuzha, securing second rank in B.Sc. (Mathematics Main) of 1973 Batch of Kerala University. It is worth mentioning that I got hundred percent marks in all Mathematics papers. My postgraduate education was in the Department of Statistics, Kariavattam Campus, University of Kerala, Thiruvananthapuram. I secured first rank in M.Sc (Statistics) of 1975 Batch. Though there was an offer of lectureship from a good college, I was interested in higher studies; hence continued in the same Department for newly started M.Phil (Statistics) course and came out in next year with A Grade.

By that time, having studied econometrics in the post graduation level, I was pondering to do research in some practical problem related to economics. Keeping this aim in view and due to other circumstances, I joined Reserve Bank of India in Mumbai on 28-09-1976 as Statistical Assistant and became Staff Officer on 28-04-1982. Then I switched over to National Bank for Agriculture and Rural Development (NABARD) and served as Assistant Manager at its Mumbai Head Office during 1983-85. With next promotion to the post of Manager, I was transferred to the regional office at Thiruvananthapuram (1985-1993) and later to Pune (1993-2001). During these twenty-five years, I could acquire professional experience by conducting various studies in connection with official work including inspection of cooperative banks.

While working in Mumbai, my intellectual life was immersed in banking and other fields of economics. I tried to do part-time research in economic department of the University of Mumbai; but my request was turned down due to lack of M. A. degree in economics. This prompted me to conduct self-study of economic texts; but I could not write exam for post graduation in this subject on account of other priorities.

Reaching Thiruvananthapuram, I continued my struggle with serious subjects alongside official work. I can recall that the seed of my interest in philosophy was sown by a special incident in 1988. On a Sunday morning, I was electrocuted while operating the washing machine in my home. Lying in water with electricity flowing through my body, unable to move or speak, I was expecting the imminent death. At that time my co-brother, who was in the next room, came and saw my danger; he rushed to put off the main switch, thus saved my life. This near-death-experience caused churning in my mind and death became a frequent subject for thought. Subsequently, the aptitude for learning philosophy had been growing in the forthcoming years. I used to get shivering due to wonder over the thought that the universe is extended infinitely without end.

Since my temperament was tuned to research activities, the job related to banking appeared to be quite routine and uninspiring. Taking books from the great libraries of British Council and Center for Development Studies in Thiruvananthapuram, I attained considerable depth in various topics of economics. The production function model of classical and neoclassical economics was the center of my intellectual fascination.

During the last period of my stay in the city, my ambition of part-time research in economics was revived. With the guidance of Professor M. A. Oommen, I prepared an article titled *Productivity of Capital Investment on Marine Fishing Crafts of Kerala* and got it published in July-Sept. 1993 issue of *Productivity*, Journal of the National Productivity council, New Delhi. On the basis of this article, I was selected in February, 1993 for research in Dept. of Economics at Kariavattam Campus of the University of Kerala. But the PhD program could not be started due to the immediate transfer of my official posting to Pune, which is reputed for better academic atmosphere. After reaching Pune, I could get the support of a guide for part-time economic research in the Mahe center of Pondicherry University, but I was denied permission on account of the lack of concerned post graduation.

Major highlight of my thought process, in the next phase, is the deviation from the ongoing study of economic themes. As a tryst with destiny, in November 1994, I purchased the popular book of Fritjof Capra, *The Turning Point*, which was originally published in 1982. Here Capra describes various layers of world – subatomic phenomena, biological organisms and different social organizations – as evolving systems, which are wholes of interconnected parts. Then he resorts to Chinese mysticism for addressing the ultimate questions. Since a whole is more than the sum of parts, it has holistic, ecological and dynamic existence. Accordingly, Capra's *systems view* of life and economy appeared to counter the mechanistic approach of the corresponding disciplines.

The treatment of systems in Capra's book triggered my critical mind and it became the turning point in my intellectual pursuit. I recognized that Capra has not explained the origin of life in the inanimate macromolecules like DNA. And, he failed to present a philosophical perspective about systems. What is the philosophy about the interconnected but layered world? How can we compare mysticism with the alternative thoughts about reality? What are the essential principles behind capitalism and communism? *These questions agitated me and I decided to take up the study of philosophy as a new venture.* I enlarged my reading by purchasing hundreds of useful books in philosophy, social sciences and related subjects from exhibitions and book stalls.

The persistence of poverty and high inequality of income-asset distribution had been a matter of grievous concern to me, in the course of analyzing the economic problems. Another issue that pained me was the cruelty and destruction due to wars as well as terrorism. Why do political leaders commit such atrocities, even though religions profess love and peace? Pondering over these problems I came to the conclusion that most of the crimes and evils are performed at the social level, rather than at individual level. The main drawback of idealism and religious philosophy is its focus on individual mind without giving due importance to the patterns of social behavior.

Thus I began to use the social perspective for deliberating about world. The most important challenge for me was to explain that social systems exist by the complementary relation of opposites. Matter-energy, space-time, body-mind, self-society, capital-labor are prominent examples of opposites, when we consider various phenomena. Continuous exposure to my books and thinking over different aspects of world generated a unique idea in me that our life is spent in seven global social systems; this principle is denoted by the phrase *seven life systems*. It became the key to start my philosophical project, which was named *System Philosophy*.

In this context, I got an intuitive idea that the *production function model* of economics would give an innovative method to study the interconnected behavior of opposite entities as well as social systems. However, for articulating this proposal, I have to study philosophical doctrines seriously. Gradually I realized that the routine of official job is a hindrance to my progress in this direction. Moreover, I was suffering from diabetes for some years on account of the continuous strain of banking career and private study. Due to the pressure of such circumstances, I voluntarily retired from NABARD service on 29-9-2001 for engaging with the research, writing and publication in the field of philosophy.

Sitting at home, I plunged into the selected books in my possession for developing the themes of System Philosophy. The dialectical method that I resorted was to take notes in English and then translate it into the mother tongue Malayalam. Then the reverse method also was adopted. This bilingual process has helped me to increase the clarity of philosophical issues.

My deep interest in the production function model of economics generated a novel idea in me for depicting the reality of universe as well as our social systems. It is the first time that a mathematical model is employed in philosophy for explaining its abstract concepts.

Inspired by the hope to articulate System Philosophy in an innovative manner, I engaged myself for three years in preparing the manuscript of my first book in Malayalam. Then I approached certain important publishers, but they refused to publish my philosophical book, holding that it is difficult to sell in current scenario. So I self-published the book in April 2004 with the title ***Saptaloka Darshanam***

Samgraham *(philosophy of seven life systems – a summary)* under the banner of PGL Books, Changanacherry, Kerala – 686 101. This work was mainly intended to explain the theoretical concepts of the seven global systems namely *nature, economy, politics, family, ethics, religion* and *art*. Due to the popular and traditional importance of number *seven*, many readers were amazed at my classification.

Next development is my participation during 2004-2005 in the program of the School of People's Economics, conducted by the NGO called VICHARA at Mavelikara, Kerala. It consisted of about thirty days of discussions and seminars on various topics, which sharpened my philosophical ideas. As a result, I prepared an article titled *From Modern Science to System Philosophy* and published it in the June 2005 issue of the journal OMEGA of ISR Aluva, Kerala.

As a matter of divine providence, I got the opportunity to join a research program during 2005-08 under the Association of Science, Society and Religion (ASSR) of Jnanadeep Vidya Peedh (Papel Seminary), Pune. The discussions and seminars conducted in December month of these years as well as the library facilities helped me to research on the interface between science, religion and philosophy. I presented a dissertation on this topic and it became the spring board for my intensive pursuit in the following years. So far I have purchased over thousand academic books that serve as authentic references for full time creative work, while remaining in the modest facilities of home.

- I presented a paper - titled *Whitehead and particle-wave duality: A critical appraisal from the perspective of System Philosophy* - in the 7th International Whitehead Conference, January 5–9, 2009 held at Dharmaram College, Bangalore.
- In March 2015, I self-published an authentic book ***jeevanum parinamavum*** (Life and Evolution) in Malayalam language discussing the theories of biological phenomena in the light of *System Philosophy*.
- I presented weekly Radio talk on various subjects regularly for six months in 2015-16.

- I have some letters to Editor regarding social issues published in news papers and periodicals of Malayalam language.
- I got an article titled *Ayurvedathinte Jnana Sithantham* (Theory of Knowledge of Ayurveda) published in the May & June 2017 issues of OUSHADHAM Journal of Ayurvedic Medicine Manufacturers Organization of India (AMMOI), Thrissur, Kerala.
- Started in 2006 a website **www.systemphilosophy.com** for presenting the topics of System Philosophy in simple English and Malayalam. I have posted a few important articles in this website.
- Further, recently I have some posts pertaining to philosophical ideas in my account at *www.facebook.com/LukeGeorge*.
- Organized the *Academy of System Philosophy* under a Trust to disseminate the new philosophy among wider audience.
- I regularly participate in seminars and discussions concerned with science, religion, philosophy and various social issues.

The foregoing is the background for preparing the manuscripts of the comprehensive books on *System Philosophy*. In the recent years I have exchanged ideas with many well placed scholars; but to my surprise they are finding it extremely hard to understand the principle of system as elaborated in my writings. Generally people are obsessed by the thought that opposite entities are separate, without any interconnection. Here I may mention the fact that acquiring knowledge is a social process, which is influenced by the ideologies and vested interests of powerful individuals of society. If a new idea comes from a lover of wisdom, who lacks the support of institutions like universities and media, it will normally face the struggle for existence. I believe that the esteemed readers of my book will help its natural selection in future because human mind has an innate tendency to prefer truth and discard falsehood.

Many teachers, friends, relatives and well wishers have helped me in the course of my life so as to contribute to the evolution of my philosophic views. My deep gratitude to all of them is beyond words. I would specifically thank Professor Hardev Singh Virk, for showing keen interest in reading my previous articles, perusing the manuscript and finally gifting me with an introduction to the present book.

I am especially indebted to many writers of philosophy and related subjects, as mentioned in the notes of chapters as well as the bibliography of this volume. I can emphasize that my limited words are not sufficient to express my acknowledgement of the ideas received from the forerunners in the history of thought.

George Luke
03 – 06 -- 2018

Introduction

Origin of Universe is the first book in the series of three books dealing with the fundamental aspects of world and phenomena. The other volumes are *Life and Mind* and *Discovery of Reality,* respectively. As a combination it forms a comprehensive treatise on REALITY by George Luke who took voluntary retirement at the age of forty eight to prepare his *magnum opus*. He is writing these books under subtitle "The Light of System Philosophy". The author has distinguished the phrase *system philosophy* from the alternative concept of *systems philosophy,* which has been popularized in the writings of Ervin Laszlo. In contrast, George Luke defines *system philosophy* as an integrative thought about the universe as a system of matter and consciousness, where these constituents are in dialectical and productive relation. This philosophical perspective about real existence involves the synthesis of rational and empirical aspects of knowledge.

The present book has been divided into seven chapters.

Chapter 1 "Fundamental Ideas of Physical Science" is an important chapter devoted to conceptual understanding of Physical Science and its evolution in Europe. The march of Physical Science from renaissance to Standard Model has been described in a systematic way. The basic principles of Physical Science are enumerated in a simple manner. The coverage is quite exhaustive but intelligible to an intelligent layman.

Chapters 2 & 3 cover Big Bang Cosmology and Quantum Cosmology, respectively. Einstein's General Theory of Relativity is in fact a theory of Cosmology. The solution of Einstein's equation leads to various models of universe. It had predicted the expansion of universe.

Different epochs of Big Bang Cosmology, experimental observations in its support, and its drawbacks are highlighted in Chapter 2. Subsequently, the author writes about Quantum Cosmology: "**Quantum cosmology is specifically concerned with the epochs constituting the physical reality. Interestingly, cosmology tries to explain the evolution of the universe up to the first second after big bang**".

Chapter 4 deals with Philosophy of Science and author defines it as that branch of philosophy dealing with the epistemology of science. The author further introduces a scheme for the scientific method in the following way: "**The true methodology of science consists of five stages, namely, *theory, hypothesis, deduction, testing* and *inductive inference*. We may introduce the phrase *TyHDTI scheme*, in order to denote scientific method conveniently**".

The opening paragraph of Chapter 5 "System Philosophy of Science" reads as follows: "In this ground-breaking chapter on philosophy of science, we will try to reconcile the conflicting doctrines of methodology, source and justification with regard to scientific knowledge. It is imperative here to consider the levels of science such as classical science, quantum mechanics, quantum field theory and quantum cosmology for the purpose of epistemological unification. Recognizing this incomplete state of philosophy of science, we would apply the principles of System Philosophy in order to derive the innovative aspects of scientific justification. The key achievement is the system model of phenomenal existence of matter and physical world".

The implications of **System Model of Justification** are described by the author: "*Our philosophical enquiry here aims to develop a theory of justification for physical science by overcoming the ambiguities and dichotomies in scientific cosmology. For achieving a breakthrough in this pursuit we must make use of the two principles, namely, principle of symmetry* and *principle of system*".

Chapter 6 "Cosmological Puzzles Finally Solved" discusses three fundamental concepts of Space, Time and Gravitational Waves. Dilemma about matter and energy has been highlighted focusing on the question of existence and failures of scientific realism. The author resorts to his system model, which gives a practical and phenomenal theory about the existence of physical world. In my view, these cosmological puzzles will remain a mystery and no final solution is possible. In the words of author: "**It is reasonable to point out that the origin of universe is an event which scientists cannot explain**".

Chapter 7 gives a summary of the topics elaborated in the two companion eBooks titled respectively as *Life and Mind* and *Discovery of Reality*. **I may appreciate the value of these separate books by mentioning the key ideas as following**.

The author focuses on Life and Evolution, firstly giving account of various theories about life, mainly genetics, and its critique. The author then refers to pitfalls of theories of evolution and writes as follows: "But many theologians and religious fundamentalists opposed Darwin's theory and evolutionary science accusing it as challenging the belief in God. Accordingly, they have published great volume of literature supporting the biblical story of creation. This aggravated the conflict between science and religion". The foremost issue in *philosophy of mind* is the definition of mind since we have to take into account the related notions like body, soul, spirit and consciousness. There are great differences between science and religion while considering the question: what is mind?" Then the author proceeds to develop the system model of human mind.

The treatise "World and Reality" is the CORE part of the volume *Discovery of Reality*. The author makes a distinction between phenomenon and reality as follows: "We can define phenomenon as any object that depends on another object through cause-effect relationship. On the other hand, **reality is the original cause of all phenomena taken as a whole**. Accordingly, reality is self-caused, infinite and permanent. The terms like *ultimate reality* and *ultimate truth* are commonly used as synonyms of reality". The bridge between reality and the set of theories of phenomenal world using the notion of worldviews is established by the author. He then introduces the **System Model of Ultimate Reality.**

Next George Luke deals with Social World and Seven Life Systems. I find this quote of author interesting: "The spectrum of social knowledge must be first divided into *scientific social knowledge* and *mystic social knowledge*, in accordance with our faculties of scientific mind and mystic mind, respectively". Then the author proposes the names of the *seven life systems*, which are global-level social systems.

For developing the "System Philosophy of God and Evil", author refers to anatomy of a crisis in the Christian world. I appreciate the idea of author to define religion as a social system: "It is a popular notion that *religion is a social system* in view of the empirical and concrete aspects like various activities of worship, the organizational structure of churches and temples as well as the social relations between the believers". Finally, George Luke starts with his definition of truth: "Truth is the property of a justified belief that it corresponds to an actual *state of affairs of the universe*". Then he discusses the dilemmas about scientific truth as well as the issues about religious truth.

Thus, adopting system approach, the two companion books can convincingly solve the age-old conceptual problems of life, evolution, mind, reality, social systems and religion.

George Luke has done a commendable job in preparing three volumes - *Origin of Universe, Life and Mind* and *Discovery of Reality* - using concepts of System Philosophy. I congratulate him for this singular achievement.

Hardev Singh Virk,

Professor of Eminence,
Punjabi University,
Patiala, Punjab (India)
www.drhsvirk.weebly.com
10 February, 2018

Chapter 1

Fundamental Ideas of Physical Science

****

1.1 Main areas of science

1.2 Classical Physical Science

1.3 Quantum Mechanics

1.4 Quantum Field Theory

1.5 Doctrines of Cosmology

Author's main original ideas are marked by [*].

The mark [#] gives the number of note at the end.

****

It is a common opinion that science advanced through the method of experiments and analysis of data. Here we distinguish science from religious faith caused by subjective experiences. As for philosophy, its original aim is to conduct critical analysis of the separate subjects like science, religion and art. This pursuit gives rise to specialized areas like philosophy of science, philosophy of religion, philosophy of art and philosophy of mind. But thinkers have not made head way in the aim to unify the different fields of our knowledge. In this context, it can be asserted that *philosophy of science* is the central branch of philosophy since it raises profound questions about the natural world as well as the conception of reality.

The framework for scientific enterprise consists of the foundational ideas such as matter, energy, space and time, which together constitute the notion of physical world. In other words, the assumption about the existence and properties of matter forms the basis for the whole edifice of science. But, due to the lack of proficiency in philosophy, the scientists as well as ordinary people are rarely aware of the validity of their accepted premises like matter exists and it is the stuff of physical world. For clearing this lacuna, it is necessary to familiarize with the fundamental ideas of science using philosophical perspective. So the present chapter aims mainly to provide an overview of the physical science developed in the last five centuries for paving the way for a systematic exposition of its philosophy. [# 1]

The scheme of this chapter may be indicated now. The first section introduces the salient aspects of science and its taxonomy. The historical development of classical physical science is given briefly in second section. Afterwards, third section is devoted to the basic components of Quantum Mechanics and in subsequent sections we will come across Quantum Field Theory and Doctrines of Cosmology.

1.1 Main Areas of Science

To begin with, it is necessary to distinguish between *content view* and *process view* about knowledge.

The descriptions about the structure of an object are often expressed in content view. It deals with the static existence of the object having such and such components. For example, consider the proposition 'earth is round'. Here 'earth' is a word under content view since it implies that earth exists statically as an object with certain perceivable components. Additionally, the word 'round' represents the geometrical object called sphere. The definition of things as appearing in a dictionary is produced under content view.

Process view is concerned with the description about the circumstances causing change in the object. Change can be alternatively viewed as an activity or process. For instance, the growth of a plant is the aspect of change happening to it; the same can be treated as the activity of plant. The growth of plant is caused by certain circumstances such as the biological features of plant as well as the external factors like availability of water, manure and proper climate. The word *context* is generally used to refer to the totality of internal and external circumstances causing change.

It may be reiterated that the process view about an object consists of the descriptions of the related context causing change. The object itself is not considered for study. When we describe the context for the growth of a plant there may be many plants satisfying such conditions. This point is better expressed by saying that the context has *multiple realisability*. To put it simply, process view aims to describe the context without considering the concerned object. Main examples of knowledge under this view include quantum physics, mysticism and theory of evolution.

We have already mentioned the two classes of propositions pertaining to the permanence and change of an object. These opposite features are complimentary. Every object has permanence or static existence for a particular period of time. And it undergoes change when we consider the static aspects of different periods. This pen has permanent existence today, but it may be destroyed tomorrow due to some circumstances. The point which we stress here is that the knowledge about an object consists of both content view and process view, both of which can be applied to fact and value.

The discussion in this book hinges on the philosophical concept called *worldview*, which is used for understanding different paradigms of physical science and other areas of knowledge. The special meaning given to *worldview* in this book may be explained now in a simple manner. It will be elaborated in the fourth chapter of present book; but a systematic treatment of this idea is postponed to the book *Discovery of Reality*.

A particular discipline of knowledge is produced by the scheme having five stages called Theory, Hypothesis, Deduction, Testing and Induction. There is large number of theories in science such as theories about subatomic particles, atoms, molecules, genetic aspects and psychological phenomena. Similarly, the various theologies of different religions and sects are developed based on the theories – basic beliefs - about God, soul and other metaphysical entities. Further, the knowledge under art has host of theories about aesthetic experience.

Here we define ***worldview*** as the set of common basic ideas found in the broadest family of theories. In other words, it contains the essential ideas of all theories having family resemblance. Hence, worldview is the method for classifying the totality of theories into the largest groups.

We arrive at the names of worldviews in two steps: Firstly, there are four worldviews namely organic worldview, mechanistic worldview, spiritual process worldview and physical process worldview. Organic worldview is the set of theories about the knowledge of *value*, while the other three worldviews deal with the knowledge of *fact* under various disciplines of science, religion and art. In the second step, it may be admitted that value can be known only through rational thinking under content view. Physical process worldview is purely empirical. But we find that both mechanistic worldview and spiritual process worldview have dual parts of rational view and empirical view. Thus there are **six worldviews** adopted in the history of human thought, which can be arranged as below.

1. Organic worldview (OWV)
2. Mechanistic worldview-rational (MWV-R)
3. Mechanistic worldview-empirical (MWV-E)
4. Spiritual process worldview- rational (SPWV-R)
5. Spiritual process worldview-empirical (SPWV-E)
6. Physical process worldview-empirical (PPWV)

The classification of six worldviews enables us to systematically draw the landscape of numerous theories pertaining to value and fact in diverse fields of knowledge. Also it is essential for comparing and contrasting the important doctrines of pioneering philosophers.

Coming to the present context, the common feature of scientific subjects may be mentioned briefly. The word **'science'** normally denotes the disciplines like physics, chemistry, astronomy, biology and psychology, which study the natural objects adopting *physical view*. Here the term 'physical' refers to matter and energy which have certain measurable properties within the framework of space and time. For example, the concepts such as weight, mass, force, energy, space, time, length and distance denote physical properties. Such properties as a whole may be alternatively called *quantitative properties*. The natural objects are normally divided into two classes, namely, inanimate things and living beings. The branch of science dealing with inanimate things is called ***physical science***; its main components are physics and chemistry. In this context, we can say that the words *physical* and *material* have the same meaning. Since mathematics and logic are abstract subjects dealing with physical world, these subjects are included in the class of physical science.

In a common sense way we treat inanimate things as physical objects. But it can be observed that living beings, especially human beings, have life and mind in addition to physical bodies. Here, mind is the collective name for the totality of mental states like ideas, emotions and desires. The characteristic features of life as well as mind are the nonphysical attributes such as creativity, purpose and freedom. Nevertheless, the project of science is to translate the nonphysical properties of life and mind into physical activities of body. This approach has resulted in the development of biology, psychology and related subjects which together constitutes the field of biological sciences. It is envisaged that the structure of biological science is built upon the foundation composed by physical science. [# 2]

Since the area of physical science has been demarcated as above, it is clear that the foundation of this subject is the principle about the existence of matter including various forms of energy. This essence of physical theory is reached through philosophical and logical thought. Scientist derives this idea from a conceptual framework designated as *worldview,* which has been introduced above. Referring to the list of six worldviews, we can state that the *mechanistic worldview* and the *physical process worldview* would together form the framework for scientific enquiry. [# 3][*]

In fact, mechanistic and physical worldviews respectively are the foundations of the two stages in the development of physical science -- Classical Science and Modern Science. Further, on the basis of historical details, it will be explained below that these stages essentially pertain to different ways of conceiving the characteristics of matter and energy.

It is instructive to give the **taxonomy of science** at this stage as per Table 1 given below. The spectrum of science is broadly divided into natural science and social science. The main branches of natural science are physical science and biological science. Alternatively, we can divide natural science into classical science and modern science. Classical science includes the fields of physics, chemistry, biology and other scientific disciplines developed during the period from sixteenth to nineteenth centuries.

In twentieth century, the principles of quantum physics were discovered. Its first phase is quantum mechanics, which was extended to the study of cosmology and biological world. The bodies of living organisms are made of cells. A cell, the basic unit of organism, is the organization of certain molecules which are ultimately decomposable to the atoms like carbon, hydrogen, oxygen and nitrogen, following the laws of quantum mechanics. Biology, psychology and related subjects, collectively called as modern biological science, are developed in this way. We may use the term **modern science** to refer to the group of scientific disciplines, which are based on quantum theory and coming under physical science and biological science.

Additionally, we have to consider the development of **social science** as the scientific study of social institutions like economy, politics and family, using the methods of mathematics and physical sciences. Social institutions originate when human beings organize their activities in particular patterns. Such institutions give special roles or meanings to participating human beings as well as to the natural things employed for the social activity. We can see the strong influence of natural science in the development of various social sciences like economics, politics and sociology. First stage of social science was developed in association with classical science, while its second stage depended on the physical process view of modern science.

Table 1 : Main Branches of Science [# 4][*]

Natural sciences	**Physical science**	Classical physical science (mainly classical physics and chemistry)
		Quantum physics – quantum mechanics and related sciences
		Cosmology (Big Bang and Quantum cosmologies)
	Biological science	Classical biology, Classical psychology and related fields
		Modern biology, modern psychology and other behavioral sciences
Social sciences	Social sciences like economics, politics and sociology.	

Physical science emerged in western countries by the combined effect of the weakening of the influence of Christian religion and the growing importance of sensory experiences in the context of studying nature. The general feature of science is to study the regularities in nature, which are commonly referred to as natural laws. Scientific pursuit aims to express the natural laws in physical terms using the properties of matter and energy – then they are called *physical laws*. This quest involves the study of nonphysical aspect of nature – life and mind – by reducing it to the physical activities of matter and energy. Accordingly, biological science, psychology and other behavioral sciences also adopt the physical approach. Hence, for practical purpose, all branches of science accept the underlying assumption that physical world and spiritual world are distinct and independent fields, involving a certain extent of dualism that will be explained in due course.

Since the disciplines of biological science and social science have been developed on the basis of physical science, we must focus on the block of physical science for enquiring into the fundamental features of science. Accordingly, it is expedient to examine the basic details of the principal parts of physical science from historical and philosophical points of view in the following sections.

1.2 Classical Physical Science

The history of science starts with the epoch of Renaissance happened in Europe in fifteenth century. The first phase of the cultural upheaval was enacted in Italy and subsequently it spread into other western countries. It is well known that Renaissance has three broad fields, namely, industrial revolution, scientific revolution and Reformation in religious organization. We may focus on the conceptual development of scientific revolution which marks the origin of science as a study of nature in physical terms.[# 5].

The scientific revolution started with the discovery of Copernicus (1473-1543) that earth and other planets revolve around the sun. Thus, Ptolemy's geocentric theory was rejected and the new heliocentric theory gained prominence. Afterwards, Kepler (1571-1630) used physical variables like place, time, distance and speed to formulate laws regarding the movement of planets. At the same time, Galileo (1564-1642), using his newly invented telescope, made significant celestial observations and conducted experiments on moving bodies. It was Galileo who articulated for the first time that science has a distinct methodology based on the physical approach of study. In this situation, Descartes (1596-1650) presented the philosophical arguments for conceiving physical world as a giant machine – this framework is called the ***mechanistic worldview***. The scientists of that period accepted the view that the world moves due to definite physical laws and hence, it is like a machine. The method of physics is to combine mathematical concepts and sensory observations to formulate the mechanical laws pertaining to nature. This mechanistic view reached its zenith when Isaac Newton (1642-1727) discovered the laws of motion and law of gravity.

In retrospect, Leucippus and Democritus of ancient Greece originally advocated the notion that the basic stuff of world is matter which consists of minute and indivisible particles called atoms. Their theory is known as *atomism*. This legacy was elaborated by Galileo, Descartes, Isaac Newton and others to establish the mechanistic worldview.

Specifically, Isaac Newton was concerned about the movement of bodies made of matter. He asserted that motion requires energy which ultimately comes from gravitational force, the origin of which was God. Obviously, Newton adopted the religious faith in creator God who caused the first motion of heavenly bodies. He envisaged that the laws of motion are uniform in all parts of space and time and it justifies the notion of mechanistic world. The so called Newtonian mechanics was highly fruitful in the study of different forms of matter and energy, leading to the rapid development of physics, chemistry and allied subjects in 18th and 19th centuries. It is customary to use the term *classical science* for referring to the science of this period. Here, classical refers to the fact that the basic tenets of Greek atomism are accepted for proposing the properties of matter in physical terms like mass, weight, length, breadth, width and so on. However, it is more appropriate to call it as *mechanistic science* for reasons given above.

The basic principles of mechanistic worldview and classical science can be outlined as following.

a) The aim of physics is to study the properties and cause-effect relations of physical world, viewed as a machine following determinate laws. Spiritual beings like God and Soul are outside the scope of physical science.

b) As per the knowledge of our scientific mind, we can treat that the physical world is real and it is made of minute particles of matter called atoms.

c) All atoms consist of homogeneous substance and it is denoted by the word *matter*. The essential property of matter is extension or mass. There are various kinds of atoms differing in size. Atoms having a particular set of chemical properties are generally called an *element*. According to this method of classification of atoms, scientists have discovered about 110 elements so far.

d) Energy is required for the motion of atoms and higher order substances. But the original source of energy is unknown to science. Newton believed that gravitational force is the basic form of energy and it is derived from God. [# 6]

e) Matter and energy are independent entities of physical world. The notion of space is intimately linked to matter while time is related to energy. In this situation space and time are absolute and independent entities.

f) The material structure formed by matter and energy is generally called as **body**. The physical world consists of bodies arranged in the levels of atoms, molecules and higher order substances. Accordingly, physical laws are the cause-effect relations between particular bodies expressed in measurable quantities.

Using the analogy of machine, the aim of classical science is to study physical world by reducing it to component parts and discovering the cause-effect relations in physical terms. This approach is called *reductionism* and it will be elaborated in chapter 4 dealing with philosophy of science.

1.3 Quantum Mechanics - Key Concepts.

The paradigm of physics underwent a radical change in the early decades of 20th century beginning with the discovery that atom has three kinds component particles, namely, proton, electron and neutron; thus it rejected the earlier theory that atom is indivisible. J. J. Thomson discovered electron in 1897. Subsequent research about the structure of atom led to the discovery of proton by Rutherford in 1914 and that of neutron by James Chadwick in 1932. The credit of suggesting the *planetary model of atom* goes to Rutherford, but his model was significantly improved later by Niels Bohr.

Thus it was established that a typical atom has a nucleus consisting of certain number of protons and neutrons. Proton has positive charge while neutron is electrically neutral. Electrons move constantly in circular orbits or shells around the nucleus, just like the planets around the Sun. This planetary model explains the three-dimensional volume of atom and also the material aspect of physical world. Electrons are negatively charged and they are equal in number to the protons so that the atom is electrically neutral. In most of the smaller atoms, the number of protons is equal to that of neutrons. But neutrons outnumber protons in most of the large atoms. In simple way, the number of protons is called atomic number; the sum of protons and neutrons is called the atomic weight. See Table 2 given below.

Table 2 : The Structure of Some Common Atoms

Atom	No. of protons (Atomic number)	No. of neutrons	No. of electrons	Atomic weight
Hydrogen	1	-	1	1
Helium	2	2	2	4
Lithium	3	4	3	7
Carbon	6	6	6	12
Oxygen	8	8	8	16
Silver	47	60	47	107
Gold	79	118	79	197
Uranium	92	146	92	238

The basic unit of three-dimensional matter is atom - the visible matter of universe is made of about 110 types of atoms. It is interesting to have an idea about the extremely small size of an atom. The diameter of hydrogen atom is 10^{-13} centimeter and the diameter of its nucleus (proton only) is 10^{-17} centimeter. Gold atom has diameter of 10^{-6} centimeter and the diameter of its nucleus is 10^{-15} centimeter. So we can say that the spherical content of an atom is mostly empty space. The size of atom is so small that it cannot be seen even by electronic microscope. A cubic centimeter of solid matter contains something like 1000,000,000,000,000,000,000,000 atoms.

Four basic forces determine the planetary model of atoms as well as the various processes within atoms and higher substances. These forces operating at the level of subatomic particles are mentioned below briefly.

First is the *gravitational force (gravity)* which is an attractive force holding the physical world as a whole -- it binds together the solar system, stars and galaxies. The magnitude of gravity is extremely small at the level of subatomic particles. The heavier the mass of a body, the stronger is the gravity.

Next is the *electromagnetic force* which operates between nucleus and surrounding electrons. It is the attractive force between positively charged protons and negatively charged electrons; neutrons are electrically neutral. The chemical reactions between atoms as well as between higher substances involve the rearrangement of the configuration of electrons and hence it is essentially related to electromagnetic force. Further this force determines the structure of molecule and the characteristics of various solids and liquids, including the properties of electricity and magnetism. The technological application of electromagnetic force is enormous. It includes the entire fields of electricity, magnetism, light and electronics (television, computers, mobile phones, etc) which involve the production and control of electrons in technical devices.

Thirdly, *strong nuclear force* is responsible for binding the protons and neutrons together in the nucleus. The breaking of nucleus, fusion of various nucleuses and the entire phenomena of nuclear energy are associated with strong nuclear force.

The fourth basic force is the *weak nuclear force* and it causes the radioactive decay of neutrons present in some heavy and unstable nuclei such as Uranium and thorium. A fundamental particle called *neutrino* is produced in a particular type of radioactive decay, that is, beta decay. In 1930, Pauli predicted the existence of neutrino, but it was experimentally detected in 1957 only. The properties of neutrino are especially interesting. This particle has very little mass, travels at the speed of light and do not interact with other particles. So it is very difficult to detect the presence of neutrinos. But through various methods, scientists have found that they are highly abundant in nature, out numbering electrons and protons by a billion to one. So there are about 55 million neutrinos per one cubic meter of space. It may be added that neutrino is the fourth subatomic particle that is stable.

The effects of strong and weak force also have great practical importance. Nuclear reactions involve the production of enormous energy; the best example is atomic bomb. Similarly, enormous energy is produced in sun and other stars by the nuclear fusion process that involve the combination of two hydrogen nuclei to form a helium nucleus along with production of energy-- this energy reaches our earth in the form of heat and light.

The realm of four subatomic particles and four basic forces can be described as the *first level of subatomic phenomena (FLSP)* because there are still deeper levels also, which will be considered later. The subatomic particles and basic forces together constitute the micro world, which explains the macro aspects of visible world consisting of atoms, molecules, higher substances, astronomical bodies and various forms of energy. In view of this fact, we treat **FLSP as the lower level of visible world**. Recollecting the points given above, we can say that the FLSP has two divisions as under:

1. *Four subatomic particles*, namely proton, neutron, electron and neutrino.
2. *Four basic forces*, namely gravitational force, electromagnetic force, strong nuclear force and weak nuclear force. The

electromagnetic, strong nuclear and weak nuclear forces together may be called as *standard forces*.

The knowledge about the properties of the above subatomic phenomena as well as their mutual relations would constitute the modern branch of physics called **quantum mechanics**; it revolutionized our knowledge about the nature of physical world made of matter and energy. We may add that the term 'quantum physics' refers to the wider discipline including the areas of quantum mechanics, quantum field theory, astronomy, cosmology and related enquiries. The two pillars of quantum mechanics are:

- Quantum theory proposed by Max Planck in 1900.
- Special theory of relativity discovered by Albert Einstein in 1905.

The path-breaking proposal of Max Planck was that the radiation of energy such as light occurs in the form of discrete particles called quanta. A quantum is the smallest unit of energy conceived as a particle. This idea was in negation of the classical view that radiation of energy travels in the form of waves. The theory of Plank is paradoxically known as the *particle property of waves* – that is, energy travels as a stream of discrete quanta. In 1905, Einstein studied experimentally the phenomenon called photoelectric effect, whereby he confirmed that light energy is not transmitted as waves, but it is a stream of quanta (packets of energy) called photons. It was the first confirmation of Plank's theory of quantum.

Einstein presented, in 1905 itself, his *special theory of relativity* which explains that particle and wave (matter and energy) are relative and inter-convertible as per the famous equation $e = mc^2$. Accordingly, matter and energy are neither independent nor absolute entities. Matter has the property of extension; therefore it needs space for existence. On the other hand, energy manifests as motion which gives rise to the concept of time. In this situation, the relativity of matter and energy can alternatively be expressed by saying that space and time are relative. It is not an exaggeration to say that the *theory of relativity* revolutionized the theoretical basis of science.

Then we see the discovery of Louis de Broglie in 1925 that a moving body behaves in certain ways as though it has wave nature. For example, a photon or any other material particle behaves like a wave having a particular wavelength. This property mentioned as *wave property of particle* is a new land mark in quantum theory. In retrospect, Thomas Young (1773-1829) and James Clerk Maxwell (1831-79) had established that light and other kinds of electromagnetic radiation have wave nature. But the implication of Broglie's discovery is that every quantum or discrete particle of radiation appears like a wave; moreover this property is a completely general one applying to all material particles. What Broglie has shown is that every material particle as well as energy quantum have the dual aspects of particle property and wave property. This phenomenon is called **particle-wave duality** and it drastically changes the concepts of matter and energy.

Consider the example of an electron. We see the particle property of electrons in the picture tube of a television set. But in some other experiments, electron appears as energy radiation which is represented by a mathematical wave function. *Hence, electron has particle (matter) property in some experiments while it has wave (energy) property in other experiments.* In this situation, the word electron does not represent a specific particle; instead it denotes a packet of energy. The most important revelation is the principle that particle-wave duality is generally observed in the case of all subatomic phenomena. It is another version of Einstein's special theory of relativity which explains that space and time are relative.

The basic idea of quantum theory is that energy radiation consists of discontinuous or discrete units of energy called quanta. It modifies the ordinary notion that a radiation is continuous flow of energy in the form of a wave. In quantum physics, energy is quantized as a stream of packets, which are similar to particles. The discovery that subatomic particles (proton, electron and neutron) have particle-wave duality has certain fundamental implications. We can conceive the macro things made of atoms through mechanistic world view.

Accordingly, it is possible to distinguish between two particles or two things; also, we see that matter and energy (space and time) are separate entities. However, subatomic world cannot be interpreted using mechanistic principles.

The intriguing aspect of subatomic phenomena is that they have the opposite properties of particle and wave. As a result the so called particles do not exist absolutely. One kind of particle can be converted into another kind of particle and also into energy. It may be added that in this context the term *wave* does not mean the visible three dimensional wave like water wave. A subatomic phenomenon is a wave in the formal sense because its existence can be expressed by a mathematical function which can be shown graphically as a wave. Here, wave represents a packet of energy. It is true that the graphical representation of energy is in the form of wave.

Next breakthrough was the discovery of Uncertainty Principle in 1927 by Werner Heisenberg, as an alternate version of particle-wave duality. The uncertainty principle means that, in the case of subatomic phenomena, the particle property is associated with uncertainty. That is, a subatomic particle does not exist absolutely as a particle, in contrast to the case of visible world. Specifically, there is an element of probability in the position of a subatomic particle. *We cannot know both the position and momentum of a particle with certainty.* When we know the position (particle aspect) correctly, there will be uncertainty about the momentum (wave aspect) and vice versa. The wave function shows the position and momentum (matter and energy) of a particle in an interrelated manner. [# 7].

The foregoing principle shows that a subatomic phenomenon – for example, an electron – actually is a particle with uncertainty. If we treat electron as a particle then it will have uncertainty of momentum. Accordingly, the position and momentum together are illustrated by a wave function involving the concept of probability. In other words, matter and energy cannot be separated as independent aspects. This is essentially what Einstein says in his principle of space-time relativity.

Particle-wave property implies that matter and energy are not absolutely different entities. They are *complementary and opposite* parts of a whole. In this situation, pioneering scientists of quantum mechanics like Werner Heisenberg, Schrödinger, David Bohm and Niels Bohr struggled hard to give a coherent explanation of physical reality. They ventured to advance philosophical narrative of the complementary and interconnected nature of matter and energy by resorting to eastern mysticism. However, we can note that the concerned scientists wanted to explain the physical process on the basis of the philosophy about immanent God under mystical perspective. [# 8].

Particle-wave duality envisages that every subatomic entity includes the properties of both particle and wave, even though there is nothing in everyday life to help us visualize that. Writers of popular books on this subject have frequently used the paradoxical description that *particle is a packet of energy*. Hence particle-wave duality is extremely puzzling; even greatest scientists could not interpret it in a coherent way. Shortly before his death, Einstein remarked that "All these fifty years of conscious brooding have brought me no nearer to the answer to the question, 'What is light quanta?'" [# 9].

In this context, **I suggest** the true meaning of particle-wave duality as following: **Every particle has inherent energy.** It means that every subatomic entity exists as a whole of material aspect and wave/energy aspect. In other words, a particle is a combination of matter and energy. When we focus on the energy aspect separately, we can describe the motion of the particle using a wave function, in which the position of particle (material aspect) involves probability. As an alternative method, we can consider the four subatomic particles together with the basic forces -- gravitational force, electromagnetic force, strong nuclear force and weak nuclear force. We can assert that the sentences like "particle is wave only", "particle is nothing but energy" and "matter is illusion" are misconceptions because it does not take into account the fact that a particle exists as a whole with the opposite properties of matter and energy. This point will be established in chapter 6, after our long journey through philosophy of science. [# 10][*].

Recall that, according to classical science following mechanistic world view, a particle or thing moves when external force is applied upon it. Newton's laws treat energy as external to the moving object.

It is natural that the principle of particle-wave duality prompted theoretical physicists to abandon mechanistic world view for studying the characteristics of subatomic particles and forces. Alternatively, they proposed the **physical process worldview** as the new paradigm for finding the laws about subatomic phenomena with particle-wave duality. It envisages to describe the activities of such phenomena, without considering the issue of existence of particular forms of matter and energy.

We may emphasize here that the discipline of *quantum mechanics* was developed on the basis of physical process worldview. It focuses on the four subatomic particles in order to study its behavior using wave functions. The variations happening in subatomic phenomena are conceived in terms of quanta of energy so that the notion of absolute particle is discarded. The components of subatomic world are not particles in the absolute sense because particle and wave are not independent entities. For example, an electron may attain different energy levels so that we cannot talk of a particular electron.

Quantum mechanics articulates the laws regarding the activities of subatomic phenomena, which are different matter-energy combinations in accordance with the theory of relativity. It can be added that quantum mechanics is like an *algorithm* that describes the interrelated activities at the subatomic level treating particles as packets of energy on the basis of the famous equation $e = mc^2$. This is in accordance with the machine-algorithm model mentioned above. The physical forms of matter and energy constitute the machine part while the pattern of activities becomes the algorithm. Since the absolute existence of particle or energy is not considered, the cause-effect relations pertain exclusively to the activities – this marks the basic deviation from the classical mechanics.

In the recent seven decades, the analysis of subatomic phenomena advanced to deeper levels consequent to two developments: (1) the discovery of the fourth particle called neutrino and (2) the principle that proton, electron, neutron and neutrino as well as the four basic forces have internal structures. Details of these new subjects will be given when we discuss the *quantum field theory* and *quantum cosmology* in next sections.

The laws of quantum mechanics are capable of explaining the physical properties and chemical activities of substances belonging to the macro level. In this way, specific phenomena such as chemical reactions, electricity, heat, light, radioactivity, gravity, electronics and atom bomb can be understood clearly. Nevertheless, the knowledge of classical science still holds good for visible level as a special case by assuming that particles exist independent of energy in tune with our sensory experience. The laws of classical science are meant to explain the physical properties and cause-effect relations, including chemical reactions, of material bodies in the macro level made of atoms.

Taking into account these facts we may postulate that the term **physical world** refers to the objects of classical science and quantum mechanics. This contention is justified by the fact that the existence of the entities of quantum cosmology causes ambiguities since these are mathematical models. Obviously, *physical world* is the world of inanimate things, conceived in physical terms.

In quantum mechanics, the focus is on the behavior of point particles since its wave property is accounted by the uncertainty principle involving probability. But, as a drawback, it could not explain the processes of basic forces. Especially quantum mechanics failed to show how the basic forces are applied between point particles. Hence the following phenomena cannot be explained adequately:

1. The change in the state of electron and the emission or absorption of photon (the quanta of light) due to electromagnetic force.
2. The travel of light through empty space as well as the effect of gravitational force at great distances. This issue is called as action-at-a-distance problem.
3. Transformation of subatomic particles from one type to another in high energy situation.

In order to solve the above problems, *Quantum Field Theory* was proposed; its details are given below in a concise manner.

1.4 Quantum Field Theory

The aim of quantum field theory is to achieve the unification of subatomic particles and basic forces. Accordingly theoretical physicists envisaged the existence of second level of subatomic phenomena (SLSP), which lies below the realm of quantum mechanics, so as to get inferences about the more elementary components of matter and energy existing in the physical world. In this new method, particles and forces are not separate entities; but are unified by the notion of *field*. It may be clarified that the term *field* is used here in a sense different from that in the context of visible world. Energy is expressed by mathematical function, the graph of which has the shape of a wave. In a sample way a packet of energy is described as the combination of many waves and thus it appears as a field. And this is the reason why a subatomic phenomenon, which is a packet of energy, is studied considering it as a field having wave function involving probability.

In the mathematical language of quantum field theory, the subatomic particles like electron and proton, which have uncertainty, are expressed as fields of energy. This is the shift from the particle property to wave property. The so called particle does not exist at a point but is spread out like a cloud of field. According to this wave description, the mass of a particle is expressed as a quantity of energy on the basis of the famous equation $e = mc^2$. To illustrate this point the mass of electron is taken to be equivalent to one unit of energy. Then mass of proton is 1836 units of energy.

Another important proposal of QFT is that the four basic forces – electromagnetic force, strong nuclear force, weak nuclear force and gravitational force – are created due to the exchange of **virtual particles.** For example, the electromagnetic force between a proton and electron is due to the exchange of particular kind of virtual particle called photon. In other words, protons and electrons interact through the exchange of photons. Consequently when an electron moves to a lower energy state, photon (light quanta) is emitted. *The notion of exchange particle (virtual particle), for representing a basic force, is the* *central feature of quantum field theory*. Thus a basic force is formally described as the exchange of particular kind of force-carrying exchange (virtual) particles. Such exchange particles are more popularly called as **bosons** or field particles.

Quantum field theory studies the interconnected behavior of various fields representing material particles and exchange particles. Then particles and forces are not independent entities. Since field is an entity spread out in space and time, it eliminates the action at a distance problem. Further, the production of unstable particles happening in atom smashers, which are technically called as ***particle accelerators***, could be explained in terms of energy fields. In this way, different elementary particles can be classified and unified.

We can find three phases in the development of quantum field theory leading to the construction of deeper models about elementary particles.

Phase 1. This is the area of *quantum electrodynamics (QED)*, which describes the interaction between electron and proton under electromagnetic force. It includes the variations in the field of electrons involving the emission and absorption of photons.

Phase 2. It consists of the laws regarding the unification of weak nuclear force and electromagnetic force pertaining to radioactive decay. This theory is alternatively called as *electroweak dynamics* because it is related to the interaction between neutron and electron within the atom, causing neutron decay. Such decay of neutron happens in free space also. Neutrino is a stable particle produced in these processes. Additionally, certain unstable particles are also produced, though they last for only less than millionth of a second; they are muon, muon-neutrino, tau and tau-neutrino. There are three exchange particles involved in the weak nuclear interaction and they are denoted as W^+, W^- and Z.

Phase 3. It is the development of *quantum chromo dynamics (QCD)* expounding the ultimate constituents of atomic nucleus.

The protons and neutrons are held together in the nucleus by the strong nuclear force caused by the exchange particles called mesons. QCD postulates that protons, neutrons, mesons and other unstable particles of strong nuclear interaction are ultimately made of quarks. Gluon is the exchange particle which binds quarks to form the said higher particles.

How did theoretical physicists propose the names of fundamental particles and why is there a tendency for increasing its number and variety? The guiding purpose is the unification of four subatomic particles and four basic forces using more elementary entities. In this pursuit, the main tool used is the mathematical theory of *symmetry*, which comes under group theory. In theoretical physics, symmetry has a precise meaning. An equation has symmetry if it remains unchanged when we shuffle or rotate its components; for example when we change space into time, time into space, electrons into quarks and so on. The notion of symmetry has great significance in cosmology also, as we will see in third chapter. [# 11].

Let us take the case of tree, for illustration. It has large number of roots in order to match the diversity for the stem, branches and leaves. Another example is a tall building which requires a wider base in comparison to that of a small building. When physicists take into account the numerous properties of subatomic particles, basic forces, atoms, molecule and higher substances, the number of fundamental particles must be sufficiently large.

The unification of electromagnetic force and weak nuclear force was established by Weinberg, Salam and Glashow in 1967 resulting in the *electroweak theory*. It was required to be combined with the Quantum Chromo Dynamics (QCD), where the latter deals with the function of quarks and gluons accounting for the strong force. In 1974, Georgi and Glashow developed the first method of grand unification of electroweak force and strong nuclear force. Subsequently, the *supersymmetry theory* was proposed by many physicists, in order to unify the various particles and forces.

The elementary constituents of a particular subatomic particle are discovered through experiments in laboratories known as *particle accelerators*. During 1950s and following decades, about 100 particle accelerators were set up in USA and Europe. The biggest and most important among them is the Large Hadron Collider (LHC) set up by 2007 at CERN in Geneva, Switzerland. Main part of this laboratory is a tunnel system called Particle Accelerator which is 27 Kilometers long. The experiment in it is as following. Particles of proton are sent from opposite directions. They are accelerated to very high speeds and made to collide with each other. As the result of collision, protons are shattered into pieces which are photographed.

The tracks of energy appearing in the photographs are later analyzed for determining their energy levels and other physical features. By classifying such tracks of energy it was discovered that protons and neutrons are made of different kinds of elementary particles called quarks. Proton is made of two up quarks and one down quark. Hence, we can say: proton = (up, up, down). Neutron is a combination of down, down and up quarks. In addition to up and down quarks there are four other types of quarks namely, charm, strange, top and bottom. The experiments of LHC revealed the existence of electron and neutrino with different varieties; these are collectively called as *leptons*.

The ultimate particles of the second level subatomic phenomena (SLSP) are neatly arranged in a scheme called **the standard model**. To regain our perspective, the standard model consists of 18 fundamental particles as below:

(i) *Six quarks* (up, down, charm, strange, top, bottom)
(ii) *Six leptons* (electron, electron-neutrino, muon, muon-neutrino, tau, tau-neutrino)
(iii) *Six bosons* (photon, gluon, W^+, W^-, Z and Higgs)

According to quantum field theory (QFT), each of the eighteen fundamental particles in the standard model is a field of energy having wave property. Alternatively, this energy field can be quantized and then visualized as a packet of energy quanta — thus we see the particle property. Some additional features of the standard model may also be mentioned. The six quarks and six leptons together are called *fermions*, because they have spin ½ or its odd integer multiples (that is, ½, 3/2, 5/2, ……). As a result they possess particle property. So fermions can be regarded as material particles. The solidity of matter and its chemical properties are caused by fermions. On the other hand, as explained earlier, bosons are exchange particles (virtual particles) that represent the four basic forces. They are field particles with integer spin (0, 1, 2, 3, ….) and they account for the aspect of energy in physical world.

There are three families of fermions (matter) in standard model. The first family of fermions constitute the four stable subatomic particles (proton, neutron, electron and neutrino), which account for the total mass (10^{50} tons) of universe. Proton is made of three quarks, namely up, up and down. Neutron is a combination of down, down and up quarks. The second and third families of fermions cause the production of unstable particles like mesons, sigma, lambda and omega – they get decayed in very short time forming energy.

We may recall that a subatomic entity has both wave property and particle property. This duality is expressed by Einstein's famous equation $e = mc^2$. Mass is an essential aspect of particle property and it is equivalent to a certain amount of energy. Then mass can be expressed in the units of energy. If unit of mass is taken as MeV/c^2, where MeV = Mega electron Volt and c = speed of light (300000 kilometers per second), then mass and energy become numerically equal.

For a particle of given energy, when mass decreases to zero then speed increases up to that of light. Two bosons, namely photon and gluon are massless particles because they travel at the speed of light. The bosons W^+ and W^- form a particle- antiparticle pair. W^+, W^- and Z bosons have significantly heavy masses and this is responsible for the very short range of the weak nuclear force. Since mass of proton is 938, the W particles have about 85 proton masses and the Z particle has 97 proton masses.

It may be added that the standard model was enlarged subsequently by increasing the number of fundamental particles. The six quarks and six leptons have antiparticles also. Further, all the quarks have three colours (red, green, blue) taking their number to 36. The analysis of strong nuclear force shows that there are 8 types of gluons. Hence, the elaborate picture of standard model consists of 61 fundamental particles as following: quarks = 36, leptons = 12, photon = 1, gluon = 8, W = 2, Z = 1 and Higgs boson = 1. In this picture, there are 13 kinds of virtual particles and they are collectively called bosons. As a result, through more detailed experiments, the number of fundamental particles has increased up to 61. The most simple and elegant form of **Standard Model** showing 18 elementary particles is presented in the next table.

There are many philosophical questions pertaining to the diversity of fundamental particles in standard model. Important ones are the following.

- The production of two categories of particles as fermions and bosons calls for an explanation. Why do the particles have different masses and other physical properties?
- It is widely claimed by theoretical physicists that standard model is perhaps the most experimentally successful theory ever proposed in the history of science. But we must deliberate epistemologically whether the components of standard model have experimental confirmation.
- Is there any advantage in the construction of standard model for addressing the crucial question about the origin and development of physical world? We can remark that the scientific pursuit for knowing about the constitution of physical world enlarges the issue of pluralism.

Table 3 : Standard Model

(The masses of particles are expressed in the unit of MeV/c² and given in brackets)

	Fermions or material particles			Bosons or field particles
QUARKS	UP (300)	Charm (1500)	Top (174000)	1. Photon (0) 2. Gluon (0) 3. W⁺ (79730) 4. W⁻ (79730) 5. Z (90986) 6. Higgs (125000)
	Down (300)	Strange (500)	Bottom (4300)	
LEPTONS	Electron-neutrino (very small)	Muon-neutrino (very small)	Tau-neutrino (very small)	
	Electron (0.511)	Muon (106)	Tau (1777)	
	I	II	III	
	Three families of matter			

We have already indicated that the fields of classical science and quantum mechanics actually constitute the physical world. Now it is justified to advance the argument that quantum field theory including standard model is the part of *cosmology* to be introduced below.

1.5 Doctrines of Cosmology

What is the latest knowledge about the origin and evolution of material world? Scientists answer this question resorting to the subject of *cosmology*, which combines experiments and observations of astronomy with mathematical models, imaginations and exaggerations. It is expedient here to define *cosmology* as the scientific study of the past history of universe, treating it as physical world made of matter. We note that scientists have proposed two most important theories of cosmology, namely **Big Bang Cosmology** and **Quantum Cosmology**. These cosmological doctrines will be elaborated in subsequent chapters for understanding its scientific and epistemological aspects.

NOTES of Chapter 1

#1. The specific references used for this chapter are: Beiser (2002), Bird (2003), Capra (1983), Capra (1992), Davies (1995), Davies (2007), Green (2005), Hawking (1995), Hawking (2011), Martin Curd and J. A. Cover (1998), Michio Kaku and Jennifer Thompson (2007), Newton (2010), Rosenberg (2000) and Tarnas (1991).

#2. The distinction between *physical* and *nonphysical* as given here is vital for a clear idea about science. In this connection, we may mention the meaning of the term *creation science*. Creationism is the view popularized by certain groups of Christianity in USA during 20th century holding that the biblical story of creation is literally true. It asserts that the hierarchy of inanimate objects and living beings has been created by God as described in the book of Genesis in the Bible. In this way, creationists oppose Darwin's theory of biological evolution. At the same time, the proponents claimed that creationism is scientific; thus they introduced the term *creation science* to refer to the tenets of creationism. It is the same as the biblical story of creation dressed as science. See Bird (2003), pages 2-8. The controversy about creation science is due to the misleading arguments about the features of science in contrast to non-science.

3. It is my original idea to classify the hitherto doctrines of philosophy of science on the basis of the list of six worldviews.

4. This table is my original idea because it illustrates the levels of philosophy of science as well as the position of philosophy of social science

5. Here, we have to distinguish between physical law and natural law (law of nature), as explained in this chapter. Since science undertakes the study of nature in physical or material terms, it results in the conflict between theism (religious philosophy) and materialism. This issue will be addressed in a later occasion.

6. Newton followed the religious ontology called *deism*, which is a version of theism, to be explained later.

7. Werner Heisenberg presented the Uncertainty Principle in 1927 using the method of matrix, specifically called matrix mechanics. But the wave function was originally proposed by Schrödinger.

8. We can consider the famous and historically important attempts for the interpretation of particle-wave duality as below.

- ❖ Bohr's principle of complementarity, popularly known as Copenhagen Interpretation, proposed in 1927.
- ❖ The cat-in-the-box paradox, which is a thought experiment published by Schrödinger in 1935.
- ❖ The EPR paradox -- it is a thought experiment developed by Einstein, Podolsky and Rosen in 1935.
- ❖ John Bell discovered a theorem, known as Bell's Theorem, which pertains to the interconnectedness of

subatomic particles. The terms like *non-local connection*, *non-causality*, and *quantum entanglement* emerged in this context.

A careful study of the above topics would reveal that these are mere figment of imagination, in tune with the philosophy of pantheism. We can expose the fallacy of this approach while developing the System Philosophy of Science in chapter 5.

9. Quoted from Beiser (2002), *Concepts of Modern Physics*, page 68.

10. Fritjof Capra (1992), *Tao of Physics*, pages 27, 65, 77-93 and 220-272. Capra's arguments are inspired by Chinese mysticism (pantheism). We can point out that Capra did not consider the distinction between phenomenon and reality. He wrongly proposes that the real aspects of body and mind are the same as the empirical processes of physical and mental phenomena. But Descartes had conceived body-mind dualism from metaphysical perspective. We will solve the issues about the existence of matter in chapters 5 and 6.

11 – Michio Kaku and Jennifer Thompson (2007) pages 26 and 99. See also page 75 where the authors claim that standard model is perhaps the most experimentally successful theory ever proposed in the history of science.

CHAPTER 2

BIG BANG COSMOLOGY

****

2.1 Standard Big Bang Theory

2.2 Inflationary Big Bang Theory

Author's main original ideas are marked by [*].

The mark [#] gives the number of note at the end.

****

With the purpose of defining matter and energy, we firstly divide the visible world into two levels, namely inanimate things and living beings. As we know through scientific way, the inanimate things have the essential property of extension or volume in space since they are made of matter and energy. Here *matter* is defined as the stuff responsible for the extension of a thing. It is the substratum of the physical properties of inanimate things such as mass, weight, length, width and depth. On the other hand, *energy* is defined as the force causing the motion and other changes of things. It is our commonsense view that matter and energy are separate entities in the visible world.

In the case of living beings like plants and animals, we observe two kinds of constituents described respectively as (1) *physical body* made of matter and energy, and (2) *nonphysical aspect* that is broadly divided into life and mind. Since every living being or organism is formed by the assembly of cells, it is a fact that each cell has both physical body and nonphysical aspect. It may be added that the physical body of a cell or organism is composed of certain kinds of inanimate things. In this situation, we use the notion of *physical world* for denoting the totality of bodies of inanimate and living entities, which are normally studied by the branches of science.

The notion about matter and energy is practically very important in our ordinary life since we think that objects like table and pen cannot exist if there is no matter and energy. For initiating the philosophical discussion of this topic, we would focus on two interconnected questions:

- What is the constitution of matter and energy?
- What is the origin of matter and energy?

The previous chapter has dealt with the fundamental ideas about the constitution of matter and energy as per classical theory and quantum theory. The latter consists of successive theories namely quantum mechanics and quantum field theory.

In this chapter we plan to consider the question about the origin and evolution of material world. For this purpose, scientists have developed the subject of *cosmology*, which combines experiments and observations of astronomy with mathematical models, imaginations and exaggerations. As per our definition, *cosmology* is the scientific study of the past history of universe, treating it as physical world made of matter and energy. Through the flourishing of theoretical physics and astronomical science in 20th century, scientists have proposed **Big Bang Cosmology** and **Quantum Cosmology** as the two most important theories of cosmology. The sections of this chapter are arranged for understanding the historical and methodological aspects of Big Bang Cosmology. The elaboration of the other doctrine -- quantum cosmology -- is postponed to the next chapter.

Keeping the above points as a background, we have to examine the essential aspects of big bang theory focusing on the question whether the event of big explosion happened actually. Based on the popular books on the subject, there are two versions of Big Bang Theory: Standard Big Bang Theory and Inflationary Big Bang Theory. [# 1]. When we review the treatment of these topics in the most popular books on cosmology, we find that the points are presented in a confusing manner without logical connection. So I have tried in the following paragraphs to give proper order in the arguments so as to clear the ambiguities afflicting the famous books.

It may be added that the numerical facts included in this book, with regard to cosmology and other topics, are secondary data in the form of best approximation. The differences between my reference books, in the case of a particular numerical fact, are ironed out for facilitating the understanding of ideas.

2.1 Standard Big Bang Theory

This version has become so popular that it is a part of our folklore. In 1922, Friedman had proposed a model of an expanding or contracting universe. Subsequently, Lemaitre in 1927 suggested that the universe started from the explosion of a *cosmic egg* having very small volume in which the total mass of universe had been compressed. Next landmark is the discovery of Edwin Hubble in 1929 that *the universe is expanding*. He estimated that the galaxies that are 100 million light years from us are moving away at the speed of about 2000 kms per second. One **light year** is equal to the distance travelled by light during the period of one year at the speed of 3,00,000 kilometers per second, and it is calculated as 9.46×10^{12} km (approximated as 10^{13} km). Hubble found that galaxies are receding away from us at increasing speed in proportion to the distance from earth. The more distant a galaxy is, the faster is its recession.

Now it is intriguing to ask: why are we not experiencing in normal life the expansion of universe? The specific answer is that the expansion happens between galaxies only; the bodies within a galaxy do not expand on account of the dominance of the basic forces within atoms. For illustrating the fact that the total size of a particular galaxy remains fixed, let us represent galaxies by the coins glued on the surface of a balloon. When we blow air into the balloon, it expands. Even though the surface of balloon grows, each coin remains fixed in size. The distance between any two points on a coin is unaffected, when the distance between two coins has increased. This model shows that a galaxy is a system with a gigantic network of physical laws. [# 2]

Einstein's General Theory of Relativity (1916) had predicted the expansion of universe. But Hubble's experimental discovery of expanding universe prompted contemporary cosmologists to go backwards in time and speculate that the universe was smaller in earlier periods. Based on this idea, George Gamow in 1946 presented a model about the origin of universe from a singular point. Space and time were zero at that point. Through a sudden explosion the space and time began to expand. This event was called *big bang* and it became immensely popular. Elementary particles of matter were formed out of energy during the first few minutes itself.

Through sophisticated experimental methods, cosmologists established that the birth of universe by big bang happened 13.70 billion years ago. Given the definition of light year, a light ray coming from the Big Bang stage of universe travels about 10^{23} km to reach us ($9.46 \times 10^{12} \times 1370 \times 10^7 = 10^{23}$ approximately); we can treat it as the radius of present universe. That is, the universe is a sphere, with three dimensions of space, which now has diameter of 27.40 billion light years. This extent can be called as *observable universe* since we can observe only things existing in three dimensions of length, breadth and depth.

Earlier attempts to get experimental confirmation of big bang event were first undertaken by cosmologists in 1960s. They predicted an *echo* or background radiation left by the original big bang that is the very hot and dense state at the beginning. Around 1965, Arno Penzias and Robert Wilson discovered the conclusive evidence of such echo on the basis of instruments set up on high altitude jet planes and balloons. [# 3]

The important astronomical studies about big bang were carried out by NASA using Cosmic Background Explorer (COBE) satellite in 1990-92, Wilkinson Microwave Anisotropy Probe (WMAP) satellite during 2001-03 and subsequent satellite-missions. It could provide a snapshot of the universe only at about 3,80,000 years after the origin. The phenomenon called cosmic microwave background (CMB) is very important in this respect. When a hot object cools, it emits thermal radiation. Similarly, CMB is the radiation due to the cooling of an extremely hot universe, which cooled from infinite temperature at the beginning to 6000 K when it reached 3,80,000 years of age. Hence CMB is regarded as the relic of big bang. The data about radiation was collected and analyzed for deriving vital information about the structure of universe in the early stage. [# 4]

Summarizing the relevant ideas found in the famous books on cosmology, we can develop the vital comments about the Standard Big Bang Theory as following.

There was no explosion. *The important question troubling all of us is whether the universe started by the explosion of a singularity point about 13.70 billion years ago?* Three kinds of criticisms are suggested below for refuting the idea of explosion.

First. We expect that, if the universe was created by an explosion, the distribution of matter in space would be uneven; more matter will lie near the centre of universe and less matter at farther places. *The data from COBE and WMAP showed that the universe has no discernible centre or edge. Universe appeared similar in all directions.* The universe at that early stage was like a sphere with diameter of 6×10^{18} kilometer. More importantly, the CMB was uniform from all parts of the universe. It means that matter is distributed evenly, thus contradicting the notion of explosion in the beginning.

Second. The point of big bang is alternatively called as *singularity*. The notion of *singularity* is derived from the reverse direction of the expansion of universe proved by Hubble's observation. Singularity is an extremely small and enormously dense state which accounts for energy only. We can naively say that matter did not exist at the singularity where space and time are treated as zero. The high concentration of energy is expressed in terms of infinitely large temperature. *Since singularity involves infinity, the big bang explosion is beyond the scientific method of knowledge.* In this situation, cosmologists hold that the mathematical model can go backwards only up to 10^{-43} seconds after big bang (ABB), which is called as Planck time. The events in the duration from 0 to 10^{-43} seconds ABB – termed as Planck length – are subject to speculation exclusively. [# 5]

Third. There is no satisfactory answer to the question: *what did exist before big bang?* That is, the source of energy concentrated in the singularity point is a mystery. Some scientists and theologians tried to answer this question by resorting to the religious faith in creator God holding that the past before the singularity point is the realm of God. [# 6]. According to the authorities of Roman Catholic Church the big bang represents the divine creation out of nothing (*ex nihilo* creation). Alternatively, process theologians advanced the faith that the big bang is the activity of immanent God. In these lines, large volume of interpretations has been produced to emphasize the religious view of equating big bang event with divine creation. But the combination of theology and science involve many drawbacks from philosophic point of view. Here arises the conflict between religion and science, which is not a satisfactory situation.

The above points indicate that the event of big explosion is a conjecture that suited the underdeveloped stage of cosmology in 1940s. *Now we can treat big bang as a simple metaphoric description of the beginning of universe implying that the universe started with very small size having infinite density and temperature.* This proposal can be termed as *refined big bang theory (RBBT).* [# 7][*]

Accepting this modification, there are certain evidences for **refined big bang** as following.

- *The abundance of hydrogen and helium.* The observed abundance of light elements - hydrogen and helium – throughout the cosmos confirms the evolutionary view. Cosmologists are successful in describing the development of baby universe, especially since 1 second ABB. Certain calculations are made about the reduction of temperature during this period, consequent to the expansion of universe. By applying Einstein's equation $e = mc^2$, scientists have established that material particles were produced by conversion of energy along with the cooling process. In the first phase after 1 second, the particles like proton, neutron and electron were produced abundantly. When the universe was just three minutes old, the smallest atoms called hydrogen and helium were formed by a process known as nucleosynthesis. Further cooling and expansion of universe for about 3,00,000 years resulted in the synthesis of higher atoms.

- *The formation of galaxies, stars and planets explained.* The formation of galaxies started when the universe was about a billion years old. During the period from 1 billion years to 9.10 billion years after Big Bang, about 100 billion (10^{11}) galaxies came into existence. It is further estimated that one galaxy contains 10^{11} stars. Milky Way is one among the galaxies and it includes our sun, which was formed when the universe was 9.10 billion years old. So the solar system containing our earth originated about 4.60 billion years ago. The older stars consist only of the basic atoms namely hydrogen and helium. But the younger stars, including sun, contain hydrogen (75%), helium (24%) and some higher elements (1%) like carbon, oxygen, nitrogen and iron. Also it is estimated that the present universe has the diameter of 10^{23} kilometers and it contains 10^{50} tons of matter.

We have noted above that the standard version of big bang theory is a myth; there is no scientific proof for the occurrence of the big explosion. But we can refine it to propose a less controversial theory - the refined big bang theory (RBBT) - about the beginning of universe from extremely small size. The refined theory agrees with the conditions of the baby universe at 1 second ABB, when the diameter is about 6,00,000 kilometers, as well as of the later phases of evolutionary development. However there are certain **drawbacks** in this refined version, which are highlighted as following.

First. The developments in quantum physics during 1950-80 showed that the three standard forces – electromagnetic, strong nuclear and weak nuclear – were unified as a single force at about 10^{-35} second after big bang. This situation is explained by the grand unified theory

(GUT). Since there is a conspicuous gap between RBBT and GUT, these theories must be linked by advanced doctrines of cosmology.

Second. Cosmologists have estimated that in the period from 0.0001 second to 1 second ABB, energy was converted into a set of fundamental particles. These phenomena are explained by the **Quantum Field Theory** described earlier. RBBT is deficient for accounting for this stage, which is before the realm of quantum mechanics. The properties of energy and particles in the situation of extremely high temperatures are governed by the principles of quantum physics under quantum field theory.

Third. Regarding the *initial conditions* of this universe, there are certain controversial issues like flatness problem, horizon problem, no boundary situation, mother-universe and quantum fluctuation. Since these topics are highly technical involving mathematical models, the details are not given here. However, it implies that we need further modification of RBBT to arrive at a better theory about the origin of our universe. [# 8].

2.2 Inflationary Big Bang Theory

In the wake of the foregoing drawbacks of RBBT, cosmologists engaged intensively for the theoretical and experimental improvements. The great breakthrough in the big bang riddle was achieved in 1980 when Alan Guth proposed the Inflation Theory. Discarding the notion of explosion, Guth conceived big bang as the beginning of gradual expansion of the universe from the singularity point. When universe was 10^{-35} second old, it underwent a *sudden expansion (inflation)* by about 10^{17} times. As a result, the diameter of universe increased from about 10^{-17} centimeter to 1 centimeter – that is, from the size of a proton to that of a grape fruit. The postulate of inflation marks the birth of ***Inflationary Big Bang Theory*** which is a great advancement over the refined big bang theory. The following table shows the diameter of our universe at various points in its evolutionary history. [# 9][*].

Table 1 : Expansion and Diameter of Universe

Sl. No.	Stage (Time after Big Bang)	Diameter of Universe
1.	Big Bang	0
2.	10^{-43} second	10^{-33} centimeter
3.	10^{-35} second (before inflation)	10^{-17} cm (size of proton)
4.	10^{-35} s (after inflation)	1cm (size of grape fruit)
5..	10^{-4} second	60 km
6.	1 second	6,00,000 km = 6×10^{10} cm
7.	3,00,000 years (3×10^5 years)	6×10^{18} km
8.	10^7 years	10^{20} km
9.	9 billion years	10^{23} km (10^{28} cm)

The COBE and WMAP satellite missions helped to give experimental confirmation for the theory of inflation. The data from these satellites have suggested that the large scale structure of universe about 3,80,000 years ABB was nearly uniform (homogenous). The data about cosmic microwave background (CMB) radiation *showed that universe appeared similar in all directions.* The universe at that early stage was like a sphere with diameter of 6×10^{18} kilometer. More importantly, the CMB was uniform from all parts of the universe. It means that matter is distributed evenly; the explanation for this fact came from the model of inflation.

By way of criticism, we can ask whether the experimental evidences from COBE and WMAP satellite missions can be taken as valid. Since the notion of big explosion is discarded in our preceding discussion, does the CMB radiation data offer any clue about the beginning of universe? It is possible that the concerned scientists designed the satellite experiments in order to get some data for presenting as proof for the emotionally appealing big bang theory. It involves the problem of theory influencing experiment so that the quality of scientific method is reduced. This is a serious issue, which is called underdetermination in philosophy of science.

Additionally, *the Inflationary Big Bang Theory deals with the baby universe only from the event of inflation onwards*. At the inflation phase, the age of universe was 10^{-35} second when its diameter grew from 10^{-17} centimeter to 1 centimeter. It leaves our crucial questions unanswered: What was the structure of universe before inflation? What caused the events of big bang and expansion? How did four subatomic particles and four basic forces originate during the period from 1 second to 10 seconds ABB?

Obviously, for explaining the experimental evidences about the expansion and inflation of universe as well as the ongoing production of elementary forms of matter and energy, the inflationary big bang cosmology is insufficient. In this way we can say that it has failed in answering the most important cosmological questions. So we have to turn to the quantum cosmology that is the subject of next chapter.

NOTES of Chapter 2

\# 1 The Selected Bibliography used for this chapter as well as for the next is given below.

Davies (1995), Davies (2007), Green (2005), Gribbin (2008), Hawking (1995), Hawking (2011), Michio Kaku and Jennifer Thompson (2007), Robert John Russell (Editor) (2004), Smolin (2008)

\# 2 This explanation of balloon and coins is taken from Green (2005), page 237

3 Michio Kaku and Jennifer Thompson (2007), page 134

4 Davies (2007), pages 19-25

#5 There are some alternative theories for conceiving the beginning state of universe, as following:

- The *steady state theory* advanced by Fred Hoyle and fellow scientists
- No-boundary proposal of Stephen Hawking and J. B. Hartle.
- Theory of Eternal chaotic inflation, which envisages large bubbles or mini universes, proposed by Andre Linde.
- Theory of time asymmetric initial conditions of Roger Penrose.

These theories are not generally accepted. The details of such controversial issues are available in Green (2005), Davies (2007), Hawking (1995) and Hawking (2011).

6 Davies (1995), pages 24, 132, 185-186.

7 The notion of refined big bang theory must replace the myth of standard big bang theory (* *this is my original idea*)

8 For controversies about the initial conditions of universe the following references are useful:

Green (2005), pages 238-249, 287-293, 432-435.

Davies (2007), pages 64, 79-84, 96, 151, 170, 183

Hawking (2011), pages 143, 145

9 The information from Davies (2007) is used to work out this table.

Chapter 3

Quantum Cosmology

****

3.1. Quantum Cosmology-3 (level III)

3.2. Quantum Cosmology-2 (level II)

   3.2.1 Quantum Gravity

   3.2.2 Inflation

   3.2.3 Symmetry Breaking and Higgs Mechanism

3.3. Quantum Cosmology- 1 (level I)

   3.3.1 Membrane Theory

   3.3.2 Multiverse Theory

   3.3.3 Dark Matter and Dark energy

3.4 Critical Summary and Outstanding Issues

Author's main original ideas are marked by [*].

The mark [#] gives the number of note at the end.

****

Through intensive research in theoretical physics and astronomy, scientists tried to overcome the abovementioned drawbacks of inflationary big bang cosmology. They resorted to the most exciting discoveries in quantum theory, especially in the branch of *Quantum Field Theory*. On the basis of these ideas, a new doctrine called **quantum cosmology** has emerged. It is the merger of Inflationary Big Bang Theory with Quantum Theory so as to provide proper knowledge about the beginning and development of our universe. The rationale for applying quantum theory to cosmology is that, in the very early stages of baby universe, its size was extremely small and temperature was infinitely high. The study of quantum stage is expected to solve the cosmological puzzle.

The famous publications on quantum cosmology can be grouped into two separate streams. First is the set of books and articles adhering to the purely scientific discussion of the topic. The second group consists of writings which aim to club the scientific topics of cosmology with the religious faith in God. In this case, there are several philosophic dilemmas about connecting God with physical world. Here our approach is to consider cosmology exclusively in scientific terms, without invoking the notion of God, and then to conduct the philosophic analysis. [# 1].

Quantum Cosmology tries to unify the diverse components of standard model for reaching at an ultimate cause of physical world. But if we examine the scientific literature on the subject, there is a plethora of theories which make a jumble. The most important notions appearing in this heap are symmetry, super symmetry, grand unification, string, super string, black hole, vacuum, Higgs mechanism, dark matter, dark energy, multiverse, branes and quantum gravity. Theoretical physicists like Stephen Hawking, Paul Davies, Lee Smolin and Brian Green are at their wits end about the proper arrangement of these notions involving sophisticated mathematics. Such mathematical models are often packaged and marketed as cosmological theories, but they do not present a coherent picture about the origin of universe.

So the primary task accomplished here is the sorting out of the disparate theoretical models for associating them to the chronological epochs of inflationary big bang theory. The quest for unification of the fundamental particles and forces belonging to the standard model will have to traverse the most amazing discoveries of theoretical physics during the recent five decades. We can anticipate that **quantum cosmology** tries to explain the main events in the early phases of physical world. Through deeper thinking and analysis, I propose to analyze quantum cosmology into three levels denoted by I, II and III. [# 2][*].

In order to appreciate our plan of discussion, the following three tables may be introduced. According to Table 1, quantum cosmology pertains to the levels I, II and III which together is called *invisible* world. On the other hand, visible world consists of micro world (levels IV) and macro world (level V). Since we can see the objects of IV and V through direct perception and experimental method, these levels together constitute the *visible/ observable/observed* world. But the theoretical entities belonging to Levels I, II and III are mathematical models only; they as a whole can be described as *invisible/ unobservable* world – this is usually termed as *physical reality* also. It is enlightening to say that **quantum cosmology is specifically concerned with the epochs constituting the physical reality. Interestingly, cosmology tries to explain the evolution of the universe up to the first second after big bang**. In this connection we may recall the table, given in last chapter, showing the size (diameter) of universe at different points of time in the past.

Table 1 - Visible and Invisible Levels of Physical World

Invisible level	I First level	**Quantum cosmology-1**: *Mysterious entities* like superstrings, membranes, multiverse, dark matter, dark energy and so on. This level I includes the entities prior to big bang. (S1)
	II Second level	**Quantum cosmology-2**: It consists of the earlier phenomena like Higgs mechanism, inflation, superstrings and quantum gravity and the event of big bang. It is the period up to 10^{-4} second after big bang (ABB). (S2, S3, S4, S5)
	III Third level	**Quantum cosmology-3**: It consists of fundamental particles of *standard model (quantum field theory)*. Period from 10^{-4} second to 1 second ABB. (S6)
Visible level	IV Fourth level	**Micro world** - Four subatomic particles and four basic forces. It originated in the cosmological period from 1 second to 10 seconds after big bang. The area of *quantum mechanics*. (S7)
	V Fifth level	**Macro world** - Substances made of atoms (inanimate substances including **astronomical bodies** as well as the bodies of living organisms). *Classical science pertains to this level only.* Macro world originated in the cosmological period from 10 second to present (13.70 billion years after big bang). (S8, S9, S10)

I have ventured to divide the total history of our physical world, which scientists call as universe, into ten cosmological stages from S1 to S10. The evolution of visible world can be depicted by stages S7, S8, S9 and S10 given in the following table 2. It may be noted that **no originality is claimed for the numerical facts presented in this chapter** because these are the best approximations based on my reference books.

Table 2 : The Evolutionary Epochs of Visible World

	Cosmological Stages	**Theories**	**Diameter of universe**
IV	S7 (1 second to 10 seconds ABB)	*Quantum mechanics* pertaining to subatomic particles and four basic forces.	6×10^6 km
V	S8 (10 sec to 3,00,000 years ABB)	Formation of Proton, Neutron, and clouds of Hydrogen and Helium	6×10^{18} km
	S9 (3,00,000 years to 9.10 billion years ABB)	a. Higher atoms are formed b. Formation of galaxies and stars. (1 billion years to 9.1 billion years)	10^{23} km
	S10 (9.10 billion years to present that is 13.7 billion years ABB)	Formation of earth and its evolution. Formation of life and living beings.	2.74×10^{23} km

It is natural that **quantum cosmology** was developed historically in the reverse order of levels marked as III, II, and I. So the following subsections are arranged in that order for elucidating the key concepts.

3.1 Quantum Cosmology-3. (Level III)

We have already given in the first chapter (section 1.4) the important points about *standard model(quantum field theory)*, which is now treated as the third level of quantum cosmology. As noted therein, there are many philosophical issues regarding the multiplicity of fundamental particles in standard model. Theoretical physicists are unfortunately ignorant and uncaring about the significance of philosophical enquiries. For the advancement of experimental research, they have formulated a plethora of new theories as the further stages of quantum cosmology. These topics are classified under *quantum cosmology-2* and *quantum cosmology-1*. We will now proceed to examine the theories of cosmology regarding the causative factors behind the fundamental particles of Standard Model.

3.2 Quantum Cosmology-2 (Level II)

It is reiterated that quantum cosmology-2 tries to explain the events leading to the original production of material particles within the premise of an expanding universe. There are four stages in the concerned period beginning from Big Bang (time t=0); these are denoted as S2, S3, S4 and S5. It may be mentioned that stage S1 pertains to the events before big bang, which will be described in the next subsection. The exact descriptions of the cosmological stages considered presently are shown in the following table. [# 3][*].

Table 3 : Framework of Quantum Cosmology-2

	Cosmological Stages	Theories	Diameter of Universe
Level II	S2 (Big Bang)	Big Bang Theory	Zero
	S3 (0 to 10^{-43} second ABB) = The first period after Big Bang	Theory about Quantum Gravity I = Superstring Theory and M theory	10^{-35} Centimeter
	S4 (10^{-43} s to 10^{-35} s ABB)	Quantum Gravity II	10^{-17} cm
	S5 (10^{-35}s to 10^{-4} second)	Inflation, Grand unified theory (GUT), vacuum at 10^{-35} second, symmetry breaking, Higgs mechanism.	1 cm to 60 kilometer.

For providing coherent explanation of the cosmological epochs, there are three inter-related concepts namely *symmetry, symmetry-breaking and unification*, which need some clarification. [# 4]

The property of **symmetry** can be easily observed in large number of natural objects. Sphere and cubic crystal are the most familiar examples. These shapes are invariant when rotated or when looked from different angles. On abstract analysis we can say that symmetry is equivalent to uniformity and simplicity.

We note that the horizontal pencil is more stable as compared to vertical one. *Hence more stability means less symmetry.* This tradeoff between symmetry and stability is an essential feature of evolution in nature. The individuals are unique or different. There is no symmetry at the level of particular individuals. But when we club the individuals according to a certain attribute, we get a class (group) which has more symmetry. The individuals belonging to a class can be changed or rotated without altering the class.

Evolution is a repetitive process of emergence of individuals out of classes. In this process of individualization, symmetry is traded for stability (individuality). The reduction in symmetry when individuals are formed out of a class is called *symmetry-breaking*. Now it is clear that the state of symmetry can be alternatively termed as *unification*. When some individuals are grouped into a class, we treat them as unified under certain criterion.

In an evolutionary framework, it is reasonable to presume that our universe started from the highest degree of symmetry and successively underwent many stages of phase transition, which are events of symmetry-breaking. As per the laws of physics, more symmetry exists in the phases with higher temperature. When water is heated above 100^0C vapour is formed, which has higher symmetry. Conversely, when temperature falls, symmetry-breaking occurs resulting in the concentration of vapour into water. The abrupt change from vapour to liquid water and then to solid ice are instances of *phase transition* invariably associated with the symmetry-breaking.

The next step is to expose the relevance of the notion of symmetry and symmetry-breaking to the subject of quantum cosmology. *Here, we must look for the phase transitions in the history of universe to know the incidences of symmetry-breaking; this is the right approach for developing quantum cosmology.* At the beginning, the universe was extremely hot because it was mostly energy confined to a minute space. Gradually it cooled causing successive events of symmetry-breaking. This resulted in the expansion of universe and formation of a hierarchy of complex physical structures. Having explained the idea of symmetry breaking, the following sections will present the theories of quantum cosmology. Let us seek to describe the precedence of higher levels of symmetry in the early history of physical world.

We include in level II the cosmological events during the period from zero to 0.0001 second after big bang. To understand the event of big bang itself, it is necessary to consider the phenomena occurred before big bang, which is included in level I to be explained in next section. The various phases of level II, considered here in chronological order, are quantum gravity, inflation, symmetry breaking and Higgs mechanism. This sequence of events resulted in the emergence of a variety of fundamental particles which are included in the Standard Model. The highlight of ensuing discussion is to explain the variation in the masses of different fermions and bosons. [# 5]

3.2.1 Quantum Gravity (the period up to 10^{-35} seconds ABB)

In the earliest phase of our physical universe, i.e. the period up to 10^{-35} second after Big Bang, the universe existed as plasma of matter and energy. The fusion of matter and energy is technically termed as **quantum gravity**. It is customary to divide this period into two stages. The first stage is the period up to 10^{-43} second. It is called as *Plank time* within which space and time cannot be separately conceived. The diameter of universe in the Plank time is estimated at less than 10^{-33} centimeter (10^{-35} meter) and it is defined as *Plank length*. The second stage of quantum gravity is the period from 10^{-43} to 10^{-35} second, wherein the universe underwent cooling as a preparation for symmetry-breaking. *The notion of Plank time is a legacy of standard big bang theory; it has no relevance in our present discussion.*

The crucial problem of quantum cosmology is to explain quantum gravity that manifests the perfect symmetry of matter and energy in the very early stage of physical world. This requires the unification of gravity (general theory of relativity) and quantum mechanics (theory of standard forces). We know that the general theory of relativity deals with big things like planets, stars and galaxies, while quantum mechanics pertains to the forces working in atoms and molecules. ***The efforts of world's greatest physicists of last century to unify gravity and quantum mechanics have not succeeded.*** It remained as a great puzzle that the two theories are incompatible. We may remark that even Albert Einstein (1879-1955) spent the last three decades of his life on a futile search for a unified field theory for combining gravitational field with electromagnetic field. In this situation, a consistent theory about quantum gravity will serve to fulfill the dream of every renowned scientist. Moreover, we have to connect quantum gravity with the framework of inflationary big bang cosmology because it takes into account the gravity-antigravity duality inherent in the universe. [# 6]

The breakthrough in the pursuit for unifying gravity and three standard forces is the ***string theory***. It is a revolutionary way of conceptualizing the first phase of physical world. Theoretical physicists adopted the term 'string' to refer to a one-dimensional strand of energy smaller than Plank length. The basic idea of string theory is that, during the stage of quantum gravity, the universe consisted of such strings. When a string gains energy, it gets stretched; when it gives up energy, it gets contracted. Additionally string can vibrate also. These features are similar to that of a rubber band.

The string theory was officially born in 1970 when Yoichiro Nambu of the University of Chicago proposed that the idea of string can replace that of the point particle. This path-breaking approach helped fellow physicists to show that the fundamental particles of standard model are expressions of the vibrations of strings. The main advantage of this method is that quarks, electrons and other particles can be conceived as vibrating strings so that the infinity problem pertaining to point particles are cleverly avoided.

When the strings belonging to the quantum gravity stage of universe vibrate in many different patterns, fundamental particles having various properties are formed. This process can be explained using the familiar example of a violin; its string can vibrate in many different ways producing a range of different sounds. Similarly, under string theory, certain patterns of vibration of string correspond to the distinct properties of exchange particles appearing in the Standard Model. The most important discovery in this context is that a particular kind of string vibration has the properties of *graviton*, which is regarded as the exchange particle pertaining to gravitational force. It is a land mark in the effort for unification of gravity and three standard forces.

Further through intensive research, by 1984 physicists have prepared the mathematical models of string vibrations corresponding to all fermions and bosons of Standard Model. The essential aspect of this advancement is the application of super symmetry method to the variety of string vibrations. Here, the term *super* indicates space-time with higher than four dimensions. The resulting symmetrical relationships between fermions and bosons constituted a new topic called *super symmetric string theory*, which is simply called as **superstring theory**. We may recall here our earlier discussion about the interrelated concepts namely symmetry, symmetry-breaking and unification. Since the state of symmetry can be treated as unification also, it is claimed that superstring theory presented a conclusive step toward the unification of particles and forces in the physical world.

Since graviton was not detected experimentally, cosmologists excluded it from the standard model. In this situation, they were worried that the unification of gravitational force and standard forces could not be achieved. In contrast, the advancement of superstring theory during 1980s was projected as the final stage of the saga of unification. Two cosmologists among the creators of superstring theory used firstly, rather mischievously, the phrase **Theory Of Everything (TOE)** in order to promote their mathematical framework. Theoretical physicists were excited by their claim that superstring theory would explain the entire spectrum of physical phenomena including the origin of universe and the emergence of different forms of material things.

To explain this theory, consider a matter of common experience. Suppose that we see an electric wire with naked eye from a distance. The wire appears as a straight line with one dimension (left / right). Now examine the wire standing close to it. If there is an ant on the wire, it can move around the wire in circular dimension also, in addition to the left/right dimension. When we speak of the wire in ant's point of view, it has a two dimensional surface, of which the length direction is part of our 3-dimensional space. To put it differently, we can say that the ant moves in four-dimensional space consisting of length, breadth, width and circularity. If we watch the wire from a distance, we see its length only and say in technical language that the extra three dimensions are curled-up or hidden.

In the context of three-dimensional space, the movement of ant on wire happens with four spatial dimensions. By extrapolating in the case of superstring, we can say that the vibrations of a superstring occur with ten spatial dimensions when the superstring itself is said to exist in nine dimensions of space. The vibrations of such superstrings produce fundamental particles and forces which are normally considered to exist in three-dimensional space, by curling up or hiding the extra six space dimensions.

But we have four reasons to be skeptical about the superstring theory, as outlined below.

- ❖ *First:* The superstring theory predicts the existence of super partners for fermions and bosons, such as selectron, squark, photino, zino, graviton, etc. These are theoretical constructs only, without any relevance for the physical universe. The multiplicity of supersymmetry particles in pairs shows a pluralist description of the basic structure of universe. Large member of parameters and constants have to be incorporated in this theory; obviously it poses a challenge to the spirit of unification.
- ❖ *Second:* As a further complication, *every supersymmetry string requires ten dimensions to exist -- nine dimensions of space and one dimension of time.* It means that the unification of gravity and standard forces is possible only if we assume six extra dimensions for space. This radical vision about quantum gravity raises a host of new theoretical issues like the geometry of nine-dimensional space.

- ❖ *Third:* Another deficiency of superstring theory is that it has five different versions, named as Type I, Type IIA, Type II B, Heterotic –O and Heterotic–E. These versions are proposed on the basis of different articulations of the extra dimensions. The fact that all the five types of superstring theory are mathematically consistent has opened new possibilities for interpretation. Now it is clear that such disparate theories do not help to realize the dream of unification.
- ❖ **Fourth**: *We can reasonably hold that pluralism and unification are antithetical.* This calls for a deeper analysis of the nature of scientific knowledge; it is the realm of philosophy to be dealt with in the next chapter. The metamorphosis of superstring theory into M-theory, which is explained below, will give further grounds to these critical comments.

Moreover, superstring theory has failed to explain the events like big bang as well as inflation and expansion of universe. Hence, though this theory was initially considered as the most potential development, it has many versions and paradoxes. Vast number of formulations about the hidden dimensions of universe is possible so that no experiment can prove or disprove the competing versions of string theory. Distressed by such failings of superstring theory, the famous theoretical physicist Lee Smolin frankly admits, in the introduction to his latest book *The Trouble with Physics,* that no new fundamental theory has been proved by experiments after the establishment of Standard Model of elementary particle physics during 1980s.[# 7].

3.2.2 Inflation (10^{-35} seconds after big bang)

In order to explain inflation we must consider the previous stage of quantum gravity as the perfect symmetry between gravity and antigravity. It existed during the period of first 10^{-35} seconds after big bang. Here, *antigravity* is defined as a positive energy, which causes the expansion of physical world. Also we may note that antigravity will be manifested later as the unified force of strong, weak and electromagnetic types; we call these forces generally as *energy*. We can interpret that in the quantum gravity stage, matter and energy were combined in the form of plasma. Then suddenly at 10^{-35} seconds the symmetry between gravity and antigravity was broken and the balance was tilted in favour of antigravity. Due to the sudden increase of antigravity force, the universe underwent **inflation**. Accordingly, the size of the universe increased by 10^{17} times and it reached a diameter of one centimeter, i.e. the size of a grape fruit. The theory of inflation was discovered by Allan Guth in 1980 and it is a landmark in the area of cosmology.

In this situation, the term **grand unified theory (GUT)** became popular. In cosmological perspective, it pertained to the stage of inflation in which three standard forces (excluding gravitational force) were united. Hence the epoch, with which we are concerned in this section, is simply called as *grand unified era*. Due to the sudden expansion of energy in the universe at the time of inflation, without any perceivable effect of gravity, it can be speculated that the universe became vacuum. Some authors like Fritjof Capra have given a mystical interpretation for the state of *vacuum* preceding the emergence of elementary particles like quarks. They have tried to give a religious view about the vacuum by linking it to the vision of ultimate reality according to Chinese and Indian mysticism. [# 8].

Initially, the most intriguing aspect of inflation theory was to explain how the material particles emerged out of the state of vacuum. In this context, cosmologists suggested the principle of **quantum fluctuations**. [# 9]. It envisages that the inflation field was subjected to random disturbances – the uncertainty principle prevailed in this situation – so that the baby universe became different from perfectly smooth. The small irregularities produced by quantum fluctuations eventually caused the production of material particles; subsequently the large structures of galaxies and stars were formed. We may note here that the principle of quantum fluctuations was later modified by the theories of Higgs mechanism, membranes and multiverse to be explained in the ensuing paragraphs.

3.2.3 Symmetry Breaking and Higgs Mechanism

Consequent to the inflationary vacuum, the universe suddenly cooled causing further symmetry breaking. In this process, the unified field got divided into two regions, namely, **Higgs Field** and *virtual particle field*. [# 10]. Higgs field is an invisible field consisting of massive virtual particles responsible for the material particle of universe. Initially VPF contains energy particles without any intrinsic mass. The interaction between Higgs field and VPF produced W and Z bosons with mass. One set of virtual particles came out of this interaction without acquiring any mass and they are identified as photon and gluon. The diversity in the production of massive and massless bosons through the so called *Higgs mechanism* is one of the mysteries of nature; it cannot be explained scientifically. Further it is believed that various types of Fermions (quarks and leptons) also emerged through Higgs mechanism during 10^{-4} to 1 second after big bang.

Attracted by the ingenuity of Higgs mechanism, theoretical physicists turned to conduct experiments for discovering the empirical properties of Higgs Field, which is supposed to be responsible for the mass of elementary particles in standard model. They proposed that Higgs field consists of a kind of quantum particle called Higgs boson and it is assumed to be originally responsible for the material aspect of world. In this situation, the authors Leon M. Lederman and Dick Teresi cleverly called Higgs boson as **'God particle'** leading to a hot controversy. [# 11].

For obtaining evidences about Higgs boson, a series of experiments involving high energy proton-antiproton collision have been conducted in the Large Hadron collider (LHC) situated at CERN in Geneva in the recent years. The latest announcement made by CERN on July 4, 2012 claims that they got evidence for the existence of Higgs boson particle with a mass of 125000 MeV/c^2, which is equivalent to 136 proton masses.

The puzzling question as to how the fundamental particles of standard model got various amounts of mass (including zero mass) is now answered resorting to Higgs mechanism. But the position of Higgs boson in relation to gravitation force raises a complicated issue that requires urgent resolution.

Gravitational force or gravity is a basic force in the first level of subatomic phenomena (FLSP) and its effect is very familiar in our ordinary life. Things fall on to ground just because earth has gravity. The movement of planets, stars and galaxies are governed by this force. The theoretical explanation of gravity is given by Einstein's General Theory of Relativity: Gravity is an attractive force pervading the entire universe and it is intimately connected with the mass of a body. The more the mass is, the stronger is the gravity. We can say that gravity exists at macroscopic level – the whole body of a thing has gravitational force.

At the subatomic level, strength of gravity is extremely small. If strength of strong nuclear force is 1, then the strength of gravity is 10^{-39} only, whereas strengths of electromagnetic force and weak force are 10^{-2} and 10^{-5} respectively. Hence gravity is not detected in the experiments about subatomic phenomena. *This is the reason why gravitational force is not included in the quantum field theory.* But some physicists talk about the gravitational field as extending the entire space and assume that it is composed of the exchange particles called *gravitons*. It is conceived without any experimental confirmation that graviton is a boson with zero mass and spin of 2, travelling at the speed of light. But graviton is excluded from the Standard Model since gravitational force lies outside the quantum field theory.

However, the exclusion of graviton from standard model is now compensated by accepting Higgs boson as the fundamental exchange particle causing the gravitational force. It is the Higgs field which gives different masses to quarks, leptons, W bosons and Z bosons. The combination of up and down quarks constitutes protons and neutrons, which form the nucleus of stable atoms. Since electron and neutrino have very small mass, the total mass of an atom, and hence that of any large body in the universe, comes from proton and neutron. *This leads* *to the conclusion that the gravitational force observable in massive bodies originate from the Higgs field and it is justified to include Higgs boson in the Standard Model. The confusion about graviton can be safely dispensed with and, at the same time, the lacuna of standard model is removed. This is the far reaching contribution of the latest LHC experiment at CERN.* [# 12][*].

We may now sum up our discussion about standard model. Due to the symmetry breaking after the end of GUT era, two opposite fields called Higgs field and virtual particle field (VPF) emerged. Higgs field is a material field composed of Higgs bosons as exchange particles. VPF is an energy field, in which the quanta do not have intrinsic mass. The interaction between Higgs bosons and energy field could produce various types of quarks and leptons as well as the exchange particles called photon, gluon, W^+, W^- and Z. This process is termed as *Higgs mechanism* and it accounts for the various masses (including zero mass of photon and gluon) of the particles presented in the above standard model. The multiplicity of fundamental particles is apt to raise bewildering questions in our mind as following:

- How can we explain the origin of Higgs mechanism, which causes the various particles with different values in the physical properties such as mass, spin and electric charge?
- Is there any justification for the phrase Theory Of Everything (TOE) to denote the synthesis of gravity and standard forces? [# 13]
- Is there a designing agency behind the production of particles with so many parameters having appropriate values?

For cracking these puzzles we have to enter into the area of philosophy and it will be taken up in next two chapters. Our immediate task is to describe the causative factors behind quantum gravity, so as to explain the big bang event.

3.3 Quantum Cosmology -1 (Level I)

(Mysterious Theories about Level I before Big Bang)

Let us concentrate again on the crucial questions as following:

- ❖ How did the universe come into existence?
- ❖ What existed before the origin of universe?
- ❖ What caused the origin and evolution of universe?

Obviously, these questions seek a coherent explanation of the factors leading to the event of big bang. A scheme for unifying the five versions of superstring theory, pertaining to quantum gravity, has been developed recently. The pioneering treatise of Edward Witten, prepared in 1995, proposed that the five superstring formulations are just like the different translations of a master theory, similar to the translations of a book into five different languages: Witten tentatively called this master theory as **M-Theory**, expressing ambiguity about its nature. There are various interpretations about what M stands for, such as membrane, multiverse, mystery, magic and mother, depending on the taste of the researcher. We use the phrase *mysterious theories* - mainly including membrane theory, multiverse theory as well as the theory of dark matter and dark energy – in order to refer to the latest cosmological doctrines.

3.3.1 Membrane Theory

As seen above, the superstring theory envisages string vibrations occurring in ten spatial dimensions. This idea is modified in *Membrane Theory*, to hold that one dimensional superstring operates upon membranes having nine spatial dimensions. More generally, a *membrane*, or *brane* for short, is defined on an object with p dimensions, where p is a whole member from 0 to 9. When p is zero, the brane is a point particle. When p is 1, brane is the same as string, a one dimensional object. When p is 2, the brane is a plane like the surface of a paper. When p is 3, the brane becomes our normal three dimensional space. It may be added that a brane with p dimension is technically called as p-brane. Accordingly the Membrane Theory postulates that the spatial aspect of universe exists as a series of 9-branes. When one dimensional superstring exists in a 9-brane, its vibrations occur with 10 spatial dimensions, as explained above.

The improvement or unification achieved in Membrane Theory is the concept of p-brane to represent the space in which superstrings operate. The value of p can go up to nine, depending on the complexity of details considered. At the level of maximum dimensions, the space of universe exists as a series of 9-branes upon which the superstrings vibrate and move. Moreover, the Membrane Theory allows for the conception of space with lower values of p, so that the extra dimensions are hidden. In this context, the term **compactification** is defined as the process of hiding certain number of higher dimensions so as to result in a simpler universe. In this manner, we live in a universe with three dimensions of space; this space consists of a series of 3-branes since the extra six dimensions are hidden.

3.3.2 Multiverse Theory

The foregoing discussion makes it clear that the quantum gravity stage of universe consists of p-branes and superstrings. So the next step is to turn to another puzzling aspect of M-theory that has stronger implication for quantum cosmology.

The nature of geometry of hidden dimensions has wide range of possibilities depending on the values of parameters. Accordingly a level of universe with 9–p hidden dimensions consists of very large number of sub-universes, having various geometries of hidden dimensions. These sub-universes are alternatively called as *pocket universes*. The total landscape of such pocket universes arranged according to the levels of hidden dimensions is named as **multiverse.**

Thus the concept of multiverse refers to the population of universes formed by various types of compactifications from 9-branes. Our present universe is just one part of the multiverse, produced through a particular compactification of 9-brane to 3-brane. Further we can show, using theoretical models, that there are many other possible *pocket universes* pertaining to 3-brane, which have distinctive modes of compactification. Additionally similar models of pocket universes – alternatively called as bubbles - can be constructed for higher branes with dimensions ranging from 4 to 9. Theoretically speaking, the innumerable pocket universes have different levels of existence, depending on the hidden dimension and various types of *compactification*. Then there is no scientific explanation about the formation of our universe with all interesting features dear to our life. Stephen Hawking, in his book *The Grand Design* (page 152), expresses his desperation as under:

> "M-Theory has solutions that allow for many different internal spaces, perhaps as many as 10^{500}, which means it allows for 10^{500} different universes, each with its own laws….. The original hope of physicists to produce a single theory explaining the apparent laws of our universe as the unique possible consequence of a few simple assumptions have to be abandoned. Where does that leave us?" [#14]

Paul Davies adopts the phrase *eternal inflation* to refer to the continuous occurrence of pocket universes, metaphorically called as bubbles also. He adds: "Eternal inflation therefore offers an inexhaustible universe-generating mechanism, of which our universe – our bubble – is but one product". [# 15]

The landscape of multiverse, with infinitely large population of pocket universes, becomes the zenith of a theoretical framework involving the notions of strings, superstrings and membranes (branes). *Moreover, it presents a fascinating vision of cosmology from the materialist perspective. Science does not admit any supernatural event which is supposed to produce our universe alone. If the origin of our universe, cast in four dimensional space-time, is to be explained without invoking the idea of God, then the multiverse theory is indispensible and cleverly contrived.* Since the quantum cosmology is in accordance with the laws of physics, it postulates the existence of multiverse so that our universe is just one among the limitless number of pocket universes.

How can we link the multiverse theory to the notion of quantum gravity, introduced at the beginning of this section? When we define quantum gravity as the first stage of our universe after Big Bang, we are focusing on a particular three-dimensional brane world produced by the compactification of 9-brane in a unique way. So in the multiverse scenario, the quantum gravity refers to the plasma of matter and energy that existed in the earliest period of an interesting pocket universe, in which we live and die. To put it differently, the idea of quantum gravity belongs to the framework of space and time used for the physical view of our pocket universe; its history is the subject of quantum cosmology.

The multiverse theory is highly speculative and it is expressed in mathematical models without any experimental evidence. So we are prone to dismiss multiverse theory as a theoretical hocus-pocus, rather than treating it as a real progress in cosmological enquiry. We have to examine the status of superstring theory and M-theory using the lens of philosophy and this task is postponed to the coming chapters. But in the recent two decades, theoretical physicists and astronomers have seen a silver lining in the horizon of quantum cosmology. Side stepping the issues of string theory and other aspects of M-Theory, the twin concepts of dark matter and dark energy have taken the centre stage in contemporary research on the origin of physical world.

3.3.3 Dark Matter and Dark Energy

There are three puzzling aspects of our universe, which provide the key to make inferences about the past universe, as following:

- Flatness of space-time
- Formation of clusters of galaxies

- Expansion of universe as a whole.

The exploration of NASA's Cosmic Background Explorer (COBE) satellite launched in 1989 and Wilkinson Microwave Anisotropy Probe (WMAP) satellite launched in 2003, established that the universe appears to be flat. So the average density of ordinary matter in our visible universe is taken as the **critical density** and it is estimated at 10^{-29} gram per cubic centimeter. It amounts to the weight of five hydrogen atoms is one cubic centimeter of space. We may further note that by ordinary matter we mean the atoms in the visible universe. It includes the energy trapped in the atoms as well as energy radiations consisting of photons which are mass less – such energy radiations arise from changes in the state of atoms.

The parts of ordinary matter have been explained using the Standard Model presented earlier. The main components of ordinary matter are leptons and quarks. In the lepton family, consisting of electrons and neutrinos, the particles have very little mass. On the other hand, quarks combine to form protons and neutrons which together constitute the nucleus of atom and hence accounts for the mass of universe estimated at 10^{50} tons. It is customary to call protons and neutrons collectively as *baryons*. So there is no problem in referring to the ordinary matter as baryonic matter, as far as the mass of visible universe is concerned. Considering the vastness of universe, we may note further that the large structures such as stars and planets in the galaxies form only one quarter of total baryonic matter, while the remaining three quarters belong to the gas existing in the space between stars and galaxies.

The geometry of universe has three possible shapes according to Einstein's general theory of relativity:

1. *Flat universe.* Here the universe exists as the surface of a plane. Then the average density of matter in the universe is called as critical density. That is, the ratio of average density to critical density (AD/CD ratio) will be equal to 1.

2. *Closed universe.* In this case universe exists as the surface of a sphere. Then the average density of matter is more than critical density. In other words, the AD/CD ratio is more than 1.
3. *Open universe.* Here the universe takes the shape of the surface of hyperbolic sheet. Then AD/CD ratio is less than 1.

A nagging problem arising from the observation of *flat universe* is: how can we reconcile it with the spherical geometry of closed universe envisaged by Einstein's general theory of relativity? Depiction of space-time as the surface of an expanding balloon, with the galaxies represented by the coins fixed on the surface, has been given in last chapter. Then what is the sense of the statement that the visible universe is flat? This question is now answered using COBE and WMAP.

Employing sophisticated techniques, scientists have calculated that the ordinary matter accounts for just 4 percent of the total matter-energy content of universe. It gives the critical density of flatness. The balance 96 percent is composed of *dark matter* (23 percent) and *dark energy* (73 Percent). *Here the adjective 'dark' indicates that such forms of matter and energy are invisible because they cannot be detected by scientific instruments. So its existence is inferred on the basis of theoretical considerations only. In this situation, the so called evidences obtained from astronomical explorations must be interpreted as indirect proofs supporting the theory.*

Adding ordinary matter to dark matter, we can say that 27 percent of total matter-energy content of universe is matter. Of course due to this much matter, the AD/CD ratio is above 1 and it explains the **spherical geometry of closed universe**, as predicted by Einstein's theory. At the same time, the visible universe made of ordinary matter appears to be flat.

We cannot visualize the shape of universe as a whole; we can imagine only about individual objects existing in the universe. Hence the notion of universe as flat, closed or open is expressed using mathematical equations and they patently come within the exclusive purview of cosmological theory. In this situation, the important question whether the shape of universe is flat, closed or open was troubling physicists for a long time. **Through the discovery of dark matter and dark energy, by 2005, this problem has been finally solved.** We can add that the structure of universe as revealed by the foregoing facts is an astonishing achievement of science and human mind. The salient aspects of dark matter and dark energy as well as the consequent interpretation of Big Bang are the topics of following discussion.

Properties of Dark Matter

Certain proofs have been suggested about the predominance of dark matter in the universe on the basis of astronomical studies regarding the clusters of galaxies and stars. Through the first billion years after Big Bang, clouds of atoms of hydrogen and helium gases were produced out of energy by a process called *nucleosynthesis*. The gravitational force operating between the atoms in the clouds of gases was too feeble to hold them together since the universe continued to expand. But the quantum fluctuations in the gas clouds caused some regions to have higher density than nearby regions. This caused the uneven distribution of matter resulting on the formation of clusters of galaxies. There were regions with different densities within each galaxy and they evolved to become stars and planets.

It is calculated that Milky Way, our galaxy, was formed when the universe was about five billion years old. The Sun and its planets were constituted after a further four billion years. *Obviously, from the uneven distribution of gases in the early universe and the subsequent formation of clusters of galaxies and stars we can infer the presence of an invisible force of gravity produced by dark matter. This is supported by the fact that the gravitational force belonging to visible matter was insignificantly small.*

In 1933, the astronomical research of Fritz Zwicky had predicted the existence of dark matter for explaining the formation of galaxies and their rotation. And the amazing advancements in the satellite technology during the recent decades confirmed the decisive role of dark matter for binding together the clusters of galaxies. Similarly, since a galaxy like our Milky Way rotates at high speed, it is the dark matter that prevents the stars from flying off from its route; otherwise the galaxy would be like an exploding flywheel.

Since the dark matter is invisible per se, the efforts to identify its physical properties can be treated as mere speculations. The most widely discussed model of dark matter is called cold dark matter (CDM), which exists in the form of super symmetry particles. The other two varieties of dark matter, namely Warm Dark Matter and Hot Dark Matter, are not to be considered here. Axion, Weakly Interacting Massive Particle (WIMP) and Neutralino are the popular candidates to be designated as the constituents of CDM. We need not go into the speculative details of these super symmetry particles as we are focusing on the role of dark matter in the history of universe.

It is now accepted that non-baryonic cold dark mater exists in the central region of galaxies, which appear as elliptical discs. In addition to different kinds of dark matter such as CDM, WDM and HDM, there is another variety of dark matter that can be qualified to be baryonic. It consists of **black holes** formed when stars die. With the complete burning of matter, the star becomes denser and denser to reach the stage of gravitational collapse. The burnt out matter of a star will have very small volume, a few kilometers of diameter at the maximum, so that its gravitational field is strong enough to bend light completely. Since light cannot escape from the collapsed star, it is invisible and hence called as black hole. Theory of black holes suggest the existence of baryonic dark matter in the regions between present stars; such an invisible stuff is called massive compact hallow object (MACHO). Interestingly, the total mass of MACHOs can be a billion times the mass of our Sun.

So far we have explained the function of dark matter in giving the distinction between flat universe and closed universe as well as causing the formation of clusters of galaxies and stars. Next task is to comprehend why the universe expanded as a whole throughout its history.

Properties of Dark Energy

It is logical to assume that there is an opposite force working against the gravity exerted by dark matter. In the absence of such an antigravity force, the entire matter of universe would clump together so that the present distribution of galaxies and stars would not be possible. The expansion of visible universe through its cosmological history is also accounted for by this antigravity force named as dark energy; the term 'dark' refers to the fact that it lies beyond the realm of our present universe. The concern here is how the existence of such mysterious energy can be explained scientifically.

As early as 1917, Einstein hit with the idea that the universe is not static, when he applied his general theory of relativity to the entire universe. A static universe would collapse by the attractive force of gravity. Hence Einstein reasoned that there is a sort of antigravity force which counterbalanced the attractive gravity and he named it as the *cosmological constant*. By 'cosmological', he meant that the antigravity force is a feature of the whole universe and not of any particular thing in it. The word 'constant' implies that the force lies behind the relativity of space-time. But Einstein was a believer in static universe – this notion was assumed by scientists of all centuries from the time of Aristotle. In this situation, Einstein contended that the function of cosmological constant was to countervail the attractive gravity, so that the universe remains static. Here, he made a mistake because in 1929 Edwin Hubble experimentally proved that the universe has been expanding after its origin at 13.7 billion years ago. [# 16]

Scientists claim that the existence of dark energy, which explains the continuous expansion of universe, is confirmed by recent astronomical observations about supernovae. When a star exhausts its nuclear fuel, it will implode under its own weight. As the star's core crashes in on itself, its temperature rapidly rises. This will result sometimes in an enormous explosion, which is called as a supernova. It is found that a single supernova explosion has the brightness of a billion suns. Such explosions of white dwarf stars are known as Type Ia supernova explosions. The astronomical studies published in 1998 about Type Ia supernovae have revealed that those explosions are caused by dark energy. And subsequently through simulation methods, the quantity of dark energy is estimated to be 73 percent of total mass/energy of the universe. The remaining 27 percent is composed of dark matter (23 percent) and ordinary matter (4 percent).

Having established the existence of dark energy as per the foregoing, astrophysicists can now address the important cosmological questions: what caused the inflation at 10^{-35} second and the subsequent expansion of universe during its history of 13.7 billion years?

The notions of inflation and expansion satisfactorily explained the apparent uniformity of our universe as revealed by COBE and WMAP satellites in the last decade. It is the preexisting opposite forces of gravity and antigravity, represented by dark matter and dark energy respectively, which caused cosmological inflation and expansion. During the first 10^{-35} second after big bang, there was perfect symmetry between gravity and antigravity. Then there occurred a sudden tilling of the balance in favor of antigravity and it resulted in the inflation of universe so as to become a sort of vacuum.

After the inflation at 10^{-35} second, the temperature of vacuum fell sharply leading to the phenomenon of spontaneous symmetry-breaking into Higgs field and virtual particle field, as explained in earlier sections. The production of material particles through Higgs mechanism increased the attractive gravity that slowed the spatial expansion pertaining to the antigravity. So the universe expanded at a moderate rate up to 7 billion years after big bang. By that time, the gravitational attraction became weak enough for the antigravity force to become dominant, and the universe entered the era of accelerated expansion. In this manner, the role of dark energy is very significant in Inflationary Big Bang theory and it has been confirmed experimentally.

3.4 Critical Summary and Outstanding Issues

As already explained, the physical world can firstly be divided into *visible world* and *invisible world*. The higher level of visible world includes the substances made of atoms (inanimate substances and bodies of living organisms) – it is denoted by level V. The physical laws about the phenomena in level V originally belong to the field of *classical science*, based on Newton's mechanics about the motion of massive bodies.

However, in the early decades of last century scientists discovered that atoms are divisible. All atoms, energy radiations and higher substances are caused by the subatomic phenomena including four subatomic particles and four basic forces. Such particles and forces have a wonderful property called particle-wave duality. The study of this level of phenomena, which is denoted by level IV, formed a new branch of physics namely the *quantum mechanics*. We may note that these phenomena are treated as visible, just because these are directly responsible for the properties of atoms and higher substances in level V.

Now we will consider the *invisible world*. In the second half of 20th century, physicists were engaged in the search for the constituents of subatomic particles and basic forces. Developing quantum field theory and suitable experiments for that purpose, it was established that the four subatomic particles and four basic forces are composed by a set of elementary particles, each of which are combinations of matter and energy. These elementary particles, involving sophisticated experiments and mathematical models, are now neatly arranged in a table called the **standard model** (level III).

Subsequently, physicists carried out intensive research to answer the question: What are the deeper phenomena that caused the formation of standard model particles? In this respect, they totally relied on mathematical models to propose the earlier phenomena like Higgs mechanism, inflation, superstrings and quantum gravity and the event of Big Bang. The set of these phenomena is appropriately marked as level II. Subsequently, theoretical physicists speculated through mathematical techniques that a host of *mysterious entities* existed before big bang; these are mainly membranes, multiverse, dark matter, dark energy - we may denote it by level I. Theoretical physicists hold that the entities of levels I, II and III represent the aspects of *physical reality*.

In order to understand the idea that the physical world has a structure with five levels, consider the example of a house. We can get the knowledge about house in five levels. First is the level of house building, viewed from outside. Second is the level of components like walls, rooms, doors and windows as shown in the building plan. Thirdly we can consider the materials used for construction of house such as bricks, iron, wood, cement, sand and water. The fourth level consists of the physical components like atoms and molecules, which constitute the building materials. As the fifth level, we can consider the subatomic particles and forces within atoms.

By definition, *cosmology* is the scientific study of the past history of physical world, especially its origin and development. From an evolutionary perspective, the foregoing stages emerged in the chronological order of the growth of physical world. It can be compared to the five levels of a tree such as roots, stem, branches, sub-branches and leaves, which occur in the order of time. Since the enquiry of cosmology is about the origin and very early stages of universe, it pertains to the *invisible world*. In this context, the latest theory about the origin of levels I, II and III constitutes the **quantum cosmology**, the details of which have been described in the beginning of this chapter. Quantum cosmology is a radical modification of the big bang theory, with two versions namely, standard big bang theory and inflationary big bang theory. Technically, we can divide quantum cosmology into three stages as below:

- **Quantum cosmology-1** :*Mysterious entities* like superstrings, membranes, multiverse, dark matter, dark energy and so on. It is level I that includes the entities prior to big bang.
- **Quantum cosmology-2**: It consists of the earlier phenomena like Higgs mechanism, inflation, superstrings and quantum gravity and the event of big bang. It is level II pertaining to the period up to 10^{-4} second after big bang.
- **Quantum cosmology-3**: The elementary particles of *standard model* (quantum field theory) produced during the period from 10^{-4} second to 1 second after big bang.

We can present below the key points of this eclectic subject, summarizing from the given details of this chapter.

- Multiverse exists eternally consisting of infinite number of pocket universes, each having 9 or above spatial dimensions. The pocket universes are made of membranes and superstrings. Our present universe originated from a particular pocket universe through a process called compactification, whereby three-dimensional space was generated by hiding the extra dimensions. This process continued throughout the history of our universe resulting in its expansion. The correct meaning of big bang is the *beginning of compactification*.
- During the first 10^{-35} second after big bang, the universe was extremely small – its diameter became only about 10^{-17} centimeter. This is the stage of quantum gravity in which matter and energy were unified to form plasma. Hence, it is the stage of perfect symmetry between matter and energy (gravity and anti-gravity).
- By the process of symmetry-breaking, a wonderful event called **inflation** occurred. The entire universe became a sort of vacuum in which only the unified standard forces existed since the effect of gravity was not manifested. This stage happened at 10^{-35} second after big bang is known as GUT stage.
- During 10^{-35} second to 10^{-4} second after big bang, further symmetry-breaking caused the formation of Higg's field and virtual particle field. Through the interaction between these opposite fields, various kinds of elementary particles of **standard model** were produced during the period from 10^{-4} second to 1 second.
- The period from 1 second to 10 second witnessed the formation of the four subatomic particles and four basic forces – this is the realm of quantum mechanics.

- The abundant formation of light nuclei, i.e. hydrogen and helium, happened during the period from 10 second to 3,00,000 years after big bang.
- The synthesis of higher atoms was followed by the formation of galaxies and stars, which happened during the period from 3,00,000 years to 9.10 billion years.
- In the stage from 9.10 billion years to 13.70 billion years (present), earth was formed and evolved to populate with living beings.

The above fascinating saga of our universe is the result of intensive research of theoretical physics undertaken in the recent three decades. However, there are certain typical and distressing comments from eminent cosmologists, which cannot be overlooked.

According to Brian Green, there is a fundamental conflict between the two major breakthroughs of 20th century physics: general relativity and quantum mechanics. We can interpret that to be the dichotomy between gravity and standard forces, which is manifested as the interconnectedness of matter and energy (particle-wave duality). It is a moot point to ask why this conflict is not resolved so far. Is it that the unified theory is beyond the scope of theoretical science? [# 17].

Lee Smolin in his latest book, *The Trouble with Physics,* expresses similar sentiment when he lists *five great unsolved problems* in theoretical physics [# 18]. We can simply mention his list as following:

- The problem of quantum gravity
- The foundational problems of quantum mechanics
- The unification of particles and forces
- The explanation of the parameters of Standard Model particles
- The mystery about dark matter and dark energy

As the final part of this summary, it is desirable to list out the glaring drawbacks of the doctrines in contemporary cosmology. But these points are to be utilized when we try to solve the puzzle of matter in chapter 6.So the list of drawbacks is postponed to that occasion. [# 19] [*].

NOTES of Chapter 3

1   A few examples of publications about cosmology from religious point of view are given below:

   a) Robert John Russell, Nancey Murphy and C. J. Isham (Editors) (1999), *Quantum Cosmology and the Laws of Nature : Scientific Perspectives of Divine Action* (Vatican Observatory Publications, Vatican City State and The Center for Theology and the Natural Sciences, Berkeley, California; second edition, 1999)
   b) The following volumes and pages of Job Kozhamthadam (Editor-in-chief), *OMEGA– Indian Journal of Science and Religion* (Institute of Science and Religion, Aluva,) are considered here

   i.   Volume 2, no.2, Dec.2003, pages 25-50
   ii. Volume 3, no.2, Dec.2004, pages 67-82
   iii. Volume 6, no.1, Dec.2007, pages 113-146
   iv. Volume 13, no.2, Dec.2014, pages 165-169

2   These cosmological tables are presented as my original idea.

3   The table showing the stages of cosmology has been prepared by synthesizing the various descriptions given in Davies (2007), Green (2005) and Gribbin (2008). (*this is my original idea*)

4   There is excellent descriptions of symmetry and symmetry-breaking in the cosmological context in:

   ■ Davies (2007), pages 159-165

- Michio Kaku, Jennifer Thompson (2007), pages 99-112
- Green (2005), pages 220-22

#5   Main references about inflation and Higgs mechanism are: Green (2005), pages 256-286; Hawking (2011), pages 143-145; Gribbin (2008), pages 48-51; Davies (2007), pages 60-64, 80-81.

#6   Detailed analysis of quantum gravity including superstring theory, black holes, membrane theory, multiverse, Dark Matter and Dark energy are given in the following references.

Green (2005), pages 294-303, 330-332, 362-424;

Davies (2007), pages 99-103;

Michio Kaku, Jennifer Thompson (2007), pages 4, 10-12, 81-98, 110-161;

Hawking (2011), pages 143-209;

#7   Smolin (2008), *The Trouble with Physics*, page viii.

#8   Capra (1992), pages 232-247

#9   Main references about quantum fluctuations: Davies (2007), pages 61-65; Hawking (2011) pages 174-177; Green (2005) pages 305-310, 329-335

#10   Green (2005) pages 254-263; Davies (2007), pages 152-166

#11   **The God Particle: If the Universe Is the Answer, What Is the Question?** is a 1993 popular science book by Nobel Prize-winning physicist Leon M. Lederman and science writer Dick Teresi. The book provides a brief history of particle physics,

Lederman explains in the book why he gave the Higgs boson the nickname "The God Particle":

"This boson is so central to the state of physics today, so crucial to our final understanding of the structure of matter, yet so elusive, that I have given it a nickname: the God Particle. Why God Particle? Two reasons. One, the publisher wouldn't let us call it the Goddamn Particle, though that might be a more appropriate title, given its villainous nature and the expense it is causing. And two, there is a connection, of sorts, to <u>another book</u>, a much older one..."

My comment is given as following: Everybody knows that the nickname 'God particle' for Higgs boson originally appeared on the cover of the book as a publishing gimmick. In fact, to say that the Higgs boson is God particle makes no sense because its position in quantum cosmology is much after the more ancestral concepts like Dark Matter, Dark energy, Big Bang, quantum gravity, superstrings and membranes (branes). Then why should we give publicity and reverence to this nick name? (**this is my original idea*)

\# 12   My proposal that Higgs boson represents gravitational force removes the problem of excluding graviton from standard model (**this is my original idea*)

\# 13   We have explained TOE later in the context of superstring theory.

\# 14   Hawking (2011), page 152

\# 15   See Davies (2007), pages 80-83, for a good description of eternal inflation.

\# 16   The cosmological constant and anti-gravity are explained in: Davies (2007), pages 54-61; and Green (2005), pages 270-286.

17   The specific reference about fundamental conflict is Green (2005), page 16. Additionally, this issue is described in Gribbin (2008), pages 30 and 34, as well as in Michio Kaku and Jennifer Thompson (2007), pages 4-11. In this context, the proposal of synthesis is mischievously called as Theory of Everything (TOE). It is enlightening to mention here that the unification of gravitation with the three standard forces essentially involves the synthesis of particle property and wave property pertaining to subatomic phenomena. This is the key problem to be addressed in fifth chapter.

18   Smolin (2008), *The Trouble with Physics*, pages 5-16.

19   The list of drawbacks given in sixth chapter is a recasting of related points found in various references.

Chapter 4

Introducing Philosophy of Science

****

4.1 What is Scientific Method?

4.2 Epistemology of Classical Science

    4.2.1 Methodology and Source

    4.2.2 Justification and Truth in Classical Science.

4.3 Epistemology of Quantum Mechanics and Big Bang Cosmology

    4.3.1 Logical Positivism

    4.3.2 Justification and Truth

4.4 Philosophy of Quantum Cosmology

    4.4.1 Methodology of Quantum Cosmology

    4.4.2 Source.

4.5 The Crisis of Philosophy of Physical Science

Author's main original ideas are marked by [*].

The mark [#] gives the number of note at the end.

****

Hereafter we adopt the convention that the word 'science' stands for natural science consisting of physical science and biological science. Then the popular definition of science is that it is a systematic study of natural things using experiments and analysis of data in order to determine the cause-effect relations. We may elaborate this notion by stating that science consists of the following stages:

- *Assuming a theory* consisting of some abstract concepts on the basis of prior knowledge.
- *The acquisition of facts or data* through experiments designed to study the phenomenon.
- *The description of facts* by definition, data analysis and classification.
- *Ascertaining the scientific laws* about cause-effect relations between different things
- *Explanation of phenomena* through the combination of theory and experiment.

Having presented a general view of the history of science and its fundamental doctrines, now we want to concentrate on the relation between science and philosophy. Specifically, it is the philosophical analysis of scientific knowledge. Thus we are concerned with the discipline called **philosophy of science** which can be defined as that branch of philosophy dealing with the epistemology of science. As a distinguishable area of thought, philosophy of science has only a short history of less than a century. The aim of this chapter is to show that it has not achieved maturity due to serious controversies related to the main areas of philosophy itself. Hence, the contents of the available reference books in this field needs thorough overhaul for developing the points of this and the following chapters. [# 1][*].

At the outset, it is worthwhile to deliberate upon the major problems to be investigated in philosophy of science. We can propose a key list as following.

1. What is the nature of the most elementary propositions of science? How are such propositions grouped in successive stages in order to arrive at the cause-effect relations? What do we mean by scientific law? How can we compare the different kinds of investigations pertaining to various branches of natural science -- physics, chemistry, cosmology, biology, psychology and so on -- with the method of social science? Is it possible to suggest that there is a typical method for all sciences? What is the scientific method? What is the difference between scientific propositions and religious beliefs in the case of natural events? The phrases like creation science, pseudoscience and parapsychology need to be explained in this context.
2. What is the scientific view about the faculties of human mind? Can scientific investigation give knowledge about unobservable entities like mind and subatomic phenomena? What is the source of the assumptions to be used in this enquiry? How are the ideas pertaining to experiments and observations are generated in human mind so as to define a scientific law?
3. How do scientists justify their discoveries? Especially, scientists adhere to the belief that matter is the substratum of all things in the physical world. Are there valid evidences to determine the existence of matter? What is the ontology or theory of reality about physical world?
4. In which way we can assess the truth of scientific laws? Is it possible to think that scientific laws hold true at all places and times? How does science progress? What is the influence of social and cultural environment on the advancement of science? There is a naïve assumption that science is objective and value-neutral. But, considering the harmful products like atom bombs and poisonous chemicals, we feel that the scientific enterprise involve the questions of ethics. Is there a way for linking scientific facts with normative values?

In this chapter we can limit the scope of philosophy of science to the field of physical science because it is the basic level of

all sciences. At later stages we will discuss the philosophy of biological science and social science separately considering the special issues in such areas.

Taking classical science in our purview, the above questions are arranged in such a way that it corresponds to the vital components of epistemology such as *methodology, source, justification* and *truth*. Systematic analysis of the methodology of scientific knowledge shows that it is a combination of deduction and induction. This principle is adopted fruitfully in classical science and modern science, which follow the mechanistic worldview and physical process worldview respectively, resulting in two branches of philosophy of science. Accordingly, there are variations in the scheme of philosophical interpretation of the scientific method depending on these contrasting worldviews. Also we will see that the difference in methodology would give rise to conflicting doctrines about source (mind), justification and truth. These issues will be discussed thoroughly in the ensuing sections to give a vivid picture of the epistemological ambiguities of science. [# 2] [*].

4.1 What is Scientific Method?

Consider the two sentences: Earth is round, Earth is yellow. We ordinarily think that the first is a scientific proposition while the second is not. The reasons for making this conclusion will be explained now systematically.

The objectives of science expressed in simple way are to observe natural phenomena in terms of physical properties, to conduct appropriate experiments and to make general inferences about cause-effect relations. In this context, we can clarify that a scientific law in the form of an equation is also a cause-effect relation. For example, consider the experiment to find the chemical formula of water, which leads to the following chemical equation: $2H_2 + O_2 = 2H_2O$. Since the atomic weight of oxygen is 16 and that of Hydrogen is 1, the above equation states that 4 grams of hydrogen and 32 grams of oxygen will give 36 grams of water. It amounts to saying that if weight of water is nine, then one part belongs to hydrogen and eight parts belong to oxygen. The proved equation can be rephrased in the form: "if hydrogen and oxygen are combined in certain proportion, then particular quantity of water will be formed".

The above example shows that a scientific law or cause-effect relation is always expressed in quantitative measures and it answers the **how** question about the phenomenon studied. The **why** question, dealing with the purpose of phenomenon, is outside the scope of science. For example see the questions: *Why do hydrogen and oxygen choose to enter into a chemical relation? Does the emergence of water display the purpose of nature?* Such questions cannot be considered in science.

Based on such experiments and inferences, the scientists try to predict future events with the intention of controlling natural phenomena. Scientific enterprise has developed wonderful equipments and programs – we can call it as *technology* that has tremendously affected whole aspects of our life. The phrase **scientific method** is commonly used to refer to the procedure of scientists for constructing the propositions about physical world. To put our discussion in a perspective, we state that scientific method deals with the following two aspects of physical world:

a) The properties of physical things
b) The regularities of physical events

The proper area of scientific method is the so called regularities which are revealed through repeated observations. Of course it includes the physical properties of various inanimate and living things. Few examples of regularities may be mentioned here. Water flows in downward direction only, not in upward direction. Iron rusts, but gold does not. Sun rises in the east and sets in the west. All crows are black. These types of regularities can be divided into two classes – *accidental regularities* and *non-accidental regularities*.

The accidental regularity is a general property of population and it is derived from the observation of a sample. The common example is: *all crows are black*. I observe as many crows as possible and find that all crows in the sample are black. It is not possible for me to observe the population of crows belonging to all parts of the world. Since I have not found a single crow which is not black, I make the inference about population that all crows are black. Another example of accidental regularity is the sentence *sun will rise tomorrow*. We have seen that sun has risen in all days in the past and it amounts to a sample only. When we say that sun will rise tomorrow, it is a statement about the entire population covering the days of past and future. It may be added here that we do not seek to find the causes in the case of accidental regularity.

The second class of regularities, which are qualified as non-accidental, are the inferences about cause-effect relations pertaining to respective populations. We can observe the cause-effect relation in a sample case of experiment or ordinary perception. Then we make an inference that the same cause-effect relation will hold good for all future experiments or observations, that is for the population.

Of course the physical world abounds in non-accidental regularities or cause-effect relations; these are the proper objects of scientific study. For example, scientists learn through repeated experiments that atmospheric humidity is the cause for the rusting of iron. This factor does not affect gold. Hence, the inference "iron rusts but gold does not" is supported by cause-effect relation; hence it is a non-accidental regularity. When we relate two things (or events) in the form of cause and effect, we derive valuable knowledge about physical world. The term *scientific method* strictly refers to the practice of scientists for reaching at the propositions about non-accidental regularities.

We can explain below that scientific method consists of five stages, called *theory, hypothesis, deduction, testing and inductive inference*. Let us denote these stages by Ty, H, D, T and I respectively. A brief description of the five components of scientific method is given below considering for example the proposition P: "*water boils at 100ºC when heated under specific atmospheric conditions*". It expresses a cause-effect relation between heating and boiling discovered through appropriate experiment. This scientific law is an inductive inference; it is formulated through five stages as explained below.

a) **Theory (Ty)** -- It may be clarified here that theory consists of the abstract concepts about the structure of physical world. Hence, theoretical concepts are derived from the worldviews adopted for scientific enquiry. Classical science follows mechanistic worldview through the empirical path as explained earlier. Its basic theory is the definition of atoms as indivisible units of matter. In contrast, the theoretical entities such as the definitions and properties of four types of subatomic particles as well as four basic forces constitute the main part of the theory of quantum mechanics. Adopting physical process worldview about subatomic phenomena, quantum mechanics is the modern framework for explaining the physical properties, chemical reactions, and other aspects of atoms and higher substances.

Additionally, theory includes the definitions or meanings of words as well as the propositions of mathematics and logic. Scientists formulate various theories for studying different aspects of physical world. Quantum mechanics, nuclear theory, relativity theory, atomic theory (planetary model of atom), kinetic theory and crystal theory are the main examples of such theories. It can be observed that physical theories can be arranged in a hierarchy of generalization. Interestingly, by 'progress of science' we mean the movement in the direction of generality; *a particular theory is explained by a more general theory.*

Physicists propose the theoretical entities in an axiomatic way based on the pre-established laws as well as earlier experiments. Hence a theoretical entity is an invisible thing as far as the proposed experiment is concerned. A theory is said to be *a priori*, because it is conceived before planning the particular experiment. [# 3].

Moreover, a theory contains at least one such invisible theoretical entity and also certain scientific laws established earlier. The theoretical entities and other abstract concepts may reflect certain preconceived notions of the scientist on account of religious faith, social customs and vested interests. Regarding the example considered above, which belongs to the area of classical science, we can take atomic theory and kinetic theory as constituting the theory (Ty) meant for explaining the motion of molecules as well as the phenomena of heat and temperature. We may recall that atoms are treated as invisible objects in the context of classical science.

> b) **Hypothesis (H)** – At least one of the theoretical entities is invisible in relation to the experiment proposed. Such invisible theoretical concepts are to be translated into observational terms; then only specific experiments would be possible. Accordingly, in the case of example P mentioned above, atoms and molecules are reduced to a set of physical properties which can be observed. Consider an example: the term 'mass' is an invisible concept in atomic theory. It is translated into 'weight' that is an observational term related to the gravity of earth. Hypothesis is framed in terms of the physical concepts expressed in observational terms and it serves as the basis for a particular experimental study. We can realize that different hypotheses are proposed for different experiments.

In the case of P: *"water boils at 100°C when heated under specific atmospheric conditions"*, the scientist already knows how to measure the visible factors of boiling such as density of liquid, heat and atmospheric pressure. Additionally, there is a law that every liquid has a particular boiling temperature in ordinary situation. So a suitable hypothesis may be that the boiling temperature of water is correlated to atmospheric pressure.

> c) **Deduction (D)** – It will be explained in the later book *Discovery of Reality* that the method of deduction involves the syllogism: premise-fact-conclusion. Certain *testable statement or equation* can be derived from the above hypothesis by applying the syllogism. Specifically, scientist reaches at a testable statement that the boiling temperature of water is 100°C. Now it is necessary to ascertain whether this statement is true on the basis of experimental evidences.

In this context, we must remember that a *testable statement or equation* is a cause-effect relation or non-accidental regularity.

d) **Testing (T)** -- In this stage the scientist designs the laboratory equipments and conduct suitable experiments. He/she collects data about observations and analyses it using the methods of mathematics and statistics. The aim of testing stage is to verify the equation proposed in the previous stage of deduction. If experimental data confirms the truth of the equation then it is accepted. In case data contradicts the deductive equation, the scientist goes back for revising the earlier assumptions and to repeat the experiment.

e) **Inductive inference (I)** -- The deductive equation has been confirmed by a sample of observations obtained in testing stage. However, the scientist has to generalize that equation to all places and times by specifying the experimental conditions – this is the process of Inductive inference. Accordingly, the proved equation is said to have universal validity. In this way, it becomes a *scientific law* pertaining to the population of instances of the phenomenon. So it is hoped that the scientific law will be proved true in all future experiments also. Here we have to address the *problem of induction* to be discussed later. A *testable statement or equation* under stage D becomes a scientific law when it is proved by experimental testing under universal conditions.

The above five steps of scientific method consists of separate types of propositions, which can be grouped into two sets as under:

- Deductive propositions (DP) including the propositions of Ty, H and D.
- Inductive propositions (IP) including the propositions of T and I.

In a more popular manner, we can say that scientific method is a combination of two ways of understanding things, namely *deduction* and *induction*. The former is the method of generating abstract ideas in the form theoretical entities, mathematics and logic to be included in the group called DP. On the other hand, induction produces IP on the basis of sensory experiences from experiments and subsequent analysis of data.

The separate stages of DP and IP can be linked as a sequence as envisaged by Carl Hempel. He proposed in 1965 that scientific method is Deductive-Nomological Model (D-N Model). It is alternatively called as Hypothetico-Deductive method. This model effectively links the two classes of propositions, namely deductive propositions (Ty, H, D) and inductive propositions (T, I). According to this treatise, we can accept that **the true methodology of science consists of five stages namely *theory, hypothesis, deduction, testing* and *inductive inference*.** We may introduce the phrase *TyHDTI scheme*, in order to denote scientific method conveniently. Then it is instructive to illustrate the scientific method as per the following diagram.

Diagram of TyHDTI Scheme [# 4] [*]

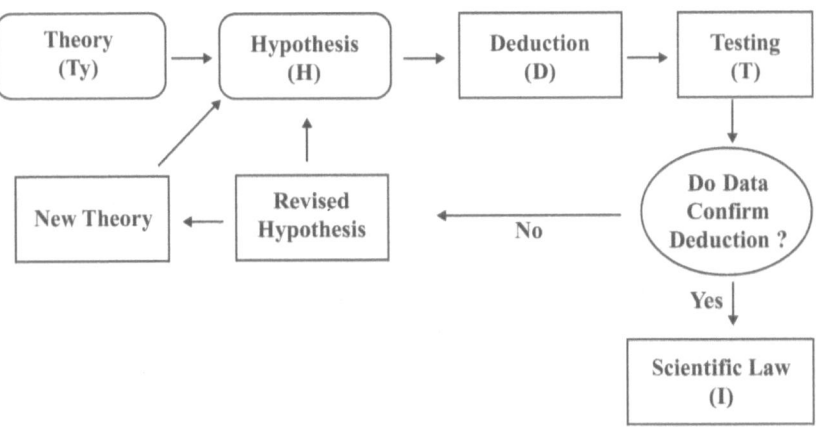

In the above diagram, there is an organic link between the various propositions used for formulating the scientific law. But the two groups of propositions -- DP and IP – are generated through separate mental processes for understanding the world, namely *deduction* and *induction*. It may be mentioned here that Hempel's D-N Model of linking DP and IP is inspired by the tenets of logical positivism, which will be explained later. Our present task is to describe the philosophical issues pertaining to the knowledge of classical science and quantum mechanics in the sequential order.

4.2 Epistemology of Classical Science

We may recall that the components of epistemology are *methodology, source, justification* and *truth*. In the heydays of classical science, following mechanistic worldview, the philosophers treated deduction and induction as separate methods without any way to link them. This situation generated the following philosophical issues: When DP and IP are two independent groups of propositions, which is the group representing the true knowledge of science? In other words, how can we assess the relative importance of DP and IP? What is the philosophy of mind regarding DP and IP? These are issues of *methodology and source*, which are inter related. In this regard, we will explain below that there are two opposite epistemological doctrines, namely, rationalism and empiricism. The question of *justification* is whether the theoretical entities represent things which have actual existence in physical world. Then only a scientific law is said to be justified. Next we must discuss the nature of *truth* applicable to scientific laws. A concise assessment of the important epistemological doctrines of classical science is presented below.

4.2.1 Methodology and Source :

Rationalism–Empiricism Conflict

The doctrine of *rationalism* holds that true knowledge is obtained only through rational thinking to produce the deductive propositions;

on the other hand, inductive inference is uncertain because it is based on sensory experience. The basic principles of rationalism were originally proposed by Descartes (1596-1650), who proposed to start with certain axioms that are *clear and distinct* ideas like definitions and other abstract concepts. Then, logical deductions are made as necessary conclusion through the method of syllogism. This method is commonly used for deriving mathematical theorems, which are regarded as absolutely certain knowledge. Taking this clue, Descartes argued that the scientific laws dealing with physical world must be formulated adopting rational method.

Descartes considered various things and its motion to conclude that the physical laws are cause-effect relations between different things made of an extended substance called matter. *He argued that matter exists really because it is defined as the substratum of the measurable physical properties like length, breadth, weight and velocity.* Such measurable aspects are classified as the primary properties of matter. The physical laws can be formulated in terms of the primary properties and it forms the subject of mathematical physics. Obviously, the properties like colour, taste and smell are not measurable; these are grouped as secondary properties, which are related to sensory experience and hence to be ignored from the purview of physical laws.

A significant idea of Descartes is that the world has real existence like a machine, similar to a giant clock, which runs according to deterministic physical laws. Thus he laid the philosophical foundations of mechanistic worldview pertaining to classical science. Since the real existence of matter, space, time, energy and cause-effect relations are deduced through abstract reasoning, this philosophy subscribes to the position of *metaphysical realism.*

The above methodology of rationalism springs from the Cartesian theory that human being possesses a metaphysical soul with the power of rational thought – this is the meaning of *cogito ergo sum* (I think, therefore I exist). Hence soul is the **source** of rational knowledge. In this situation, Descartes argues that the other aspects of human mind, including the emotions, desires and sensory experiences come under the sphere of mechanical body. The experimental data and inductive inferences are based on sense organs and hence are not produced by rational soul. The underlying *mind-body dualism* is the distinctive feature of his rational philosophy of mind; it will be discussed in the next book.

Now we may take up the aspects of empiricism, which argues that all kinds of scientific propositions are produced by inductive method. The period of hundred years from 1650 to 1750 witnessed profound change in the intellectual climate of Europe and it is known as the Age of Enlightenment. The principle features of Enlightenment can be identified as the following:

- The predominance of *physical view* upon the various aspects of nature, including manmade social systems like politics and economy.
- Reliance on *secular* and scientific method for the study of human mind, removing the shackles of religious dogma. This resulted in the new definition of *rational* as the method of combining logic with sensory data.
- As a consequence, the philosophical outlook underwent a radical shift from metaphysics to *empirical study* of nature and human reason. Philosophers began to ask: if the world is governed by physical laws, why is this not also true of human mind as well?

The empiricism of **David Hume** is the main doctrine in this context. Accepting the *ontological* position of naturalism, which was the extension of materialism from inanimate world to biological world, Hume held that natural phenomena can be studied physically using empirical data, avoiding all assumptions about supernatural forces. All natural objects are ultimately made of matter only. Accordingly Hume proposes that scientific laws are exclusively inductive inferences based on sensory data. The *methodology* of empiricism envisages giving primary importance to inductive propositions asserting that the so called deductive propositions are also derived from experimental methods. The deductive propositions including definitions, logic, mathematics and other abstract concepts are treated as the abbreviations of sensory data and they are used as the premises for induction. Hence the methodology of empiricism entails the *experimental syllogism*: premise (DP) – Testing (T) -- inductive inferences (I).

The core aspect of Hume's philosophy is that scientific laws are formed through inductive method. It involves the inference of a general principle about population, on the basis of the regularities observed for a sample. Since such regularity belongs to the sample it can be expressed as "the individuals in the sample have property K". Then, the scientist adopts the inference that "all individuals in the population have the property K". This generalization from *some* to *all* is the aim of inductive procedure. According to Hume, there is no special importance to the deductive propositions included in the theory. The abstract concepts of theory are obtained from previous knowledge about the empirical properties of things.

Regarding the *source* of scientific knowledge, David Hume adopted *epiphenomenalism* which says that the mental states such as ideas, emotions and will are the byproducts of the chemical processes in brain and nervous system. It means that mind is an epiphenomenon of matter. This view clearly denies the Cartesian body-mind dualism so that it is possible to observe mental activities in terms of external movements of body. Scientific laws or cause-effect relations are beliefs generated in human mind due to the *habit of association of ideas*. Sensory organs and brain work according to physical principles in order to become the source for scientific knowledge. In this way, Hume argues that human mind can know only the phenomenal aspects of physical world as it affect our sensory organs. Our past experiences of world have prompted us to conceive certain abbreviations. The idea of matter is one such abbreviation which represents the repeated observation of extended things.

The most important criticism about inductive method is that it does not give any certain knowledge about the physical world. This issue is popularly known as **problem of induction,** which was originally articulated by David Hume. Inductive inferences are based on limited number of observations of sample. There is a possibility that the generalization from sample to population may not hold good in future. Some future evidence may refute the scientific law leading to its rejection. In this situation, the inductive laws do not have any logical necessity; they are mere conception of *accidental regularities* as in the case of sentences like "all crowds are black" and "sun will rise tomorrow". This view obviously expresses skepticism about the existence of physical laws.

Induction does not allow us to proceed from observed regularities to stable *laws* of nature. The scientific laws are treated as a sort of opinions or constructions of mind, without representing the objective aspects of external world. We may state that the materialist approach underlying *epiphenomenalism* cannot account for the self-consciousness of our mental states like ideas, emotions and desires. We can note that the essential properties of self-consciousness are creativity, freedom and purpose. In view of this fact, we can challenge the methodological arguments of empiricism. Further it is proper to admit that the formulation of deductive propositions (Ty, H, D) display the intuition and creativity of scientists.

In the wake of the above conflict between rationalism and empiricism, **Immanuel Kant's** critical philosophy came up as a serious development. The concerned details are given in the third volume; however, we can pick out the following key points:

- Our scientific mind has a rational structure with creativity and it can be compared to a factory. Sensory data constitute the raw material for this factory which produces the outputs – firstly the deductive propositions (DP) and secondly the inductive propositions (IP).
- Though scientific propositions under both DP and IP are logically sound, these are the constructions of our scientific mind and hence are phenomenal. The knowledge about matter, physical laws and mind (self) are phenomenal ideas. Hence, the reality of universe cannot be known scientifically.

But Kant failed to account for the relative difference between DP and IP. Also he could not bridge the gap between *necessary* and *contingent truths*, pertaining to DP and IP respectively. Thus the conflict between rationalism and empiricism continues. In this situation, the exponents of empiricism wanted to strengthen the position that experimental method is the defining characteristic of science.

In the 19th century, Auguste Comte (1798-1857) and Ernst Mach (1838-1916) modified empiricism into the philosophy of **positivism**. According to this approach, the basic job of science is the description of the regularities of nature, without assuming any hypothesis about a hidden casual structure. We can note that positivists relied positively upon the experiments and evidences in order to formulate theories and scientific laws. This amounts to hostility towards unobservable aspects such as human mind and subatomic world. In this manner, the positivists proposed their arguments to hold that inductive method is reasonable for all practical purposes. The influence of positivism was more pervasive in the development of experimental psychology and social sciences.

Similarly, the techniques of statistics and probability are increasingly used for framing scientific laws about a population. For example, consider the laws, "smoking causes cancer in sixty percent of cases". Such statistical laws are based on the observation of large instances of cancer patients and the analysis of various causes for cancer. We can emphasize that this kind of causal relations are not supported by any theory about the fundamental structure of human body. However, the statistical method invariably makes use of some theory that consists of preexisting empirical laws and assumptions.

In this context, Bayes' theorem is a particularly important tool for articulating hypothesis and subsequent stages of induction. There is a doctrine called *Bayesianism* which holds that a person's degree of belief in a hypothesis would change in the light of evidence; hence, it is claimed that inductive practices can lead to truth. In the recent period, the statistical approach to induction has gained acceptability by proceeding in the path of positivism. The issue of skepticism looms large upon the empirical methods of science.

As it is admitted now, the conflict between rationalism and empiricism is a serious problem, which defies solution so far. Further, we can show that all versions of empiricism have very serious difficulties with regard to the next stages of epistemology.

4.2.2 Justification and Truth in Classical Science

Justification is the first stage of determining the validity of a proposition on the basis of evidences, such as mental images and sense data, obtained through proper methodology. It is essential to deal with this issue by taking into account the distinction between the methods of induction and deduction. Truth is defined as the second stage related to validity of propositions. So this topic will have to be discussed in the context of classical science, quantum mechanics and quantum cosmology separately. But we have to unify the different kinds of justification and truth for taking physical world as a whole. For that purpose, it is expedient to take up the discussion of scientific justification at one stretch; and it is postponed to section 5.3 of next chapter. Further, the comprehensive treatment of truth will be presented in a later stage.

4.3 Epistemology of Quantum Mechanics and Big Bang Cosmology

The subatomic phenomena like proton, neutron and electron as well as the four basic forces have particle-wave duality, making their existence ambiguous. We cannot directly observe these phenomena, but we can clearly see the effects they produce on visible objects. The most familiar example is electron, the presence of which is inferred from the effects on the picture tube of TV. It is in this sense that we treat the realm of subatomic phenomena as the lower level of visible world and its scientific laws come under the discipline called **quantum mechanics**. (Recall section 1.3 and Table 1 of chapter 3). Due to particle-wave duality, subatomic phenomena do not support the mechanistic worldview. Hence alternatively, ***physical process world view*** is adopted. It requires a new approach of scientific method. This is the background of the methodological doctrine called ***logical positivism***. In passing, we may note that a group of thinkers known as Vienna Circle originally proposed logical positivism during the period 1924-33. The prominent members of this group were Mortiz Schlick (1882-1936), Carl Hempel (1905-1997) and A.J. Ayer (1910-1989). A brief exposition of the process methodology may be given now.

4.3.1 Logical Positivism

On account of particle-wave duality, the names of objects belonging to subatomic world do not have representational property. For example, we do not hold that the word *electron* represents a specific particle. Alternatively such words denoting the subatomic phenomena are treated as *theoretical entities*. The activities of these phenomena are indirectly observed in experiments. Such observations are called evidences. It is possible to give meaning to the theoretical entity on the basis of the concerned evidence. In this context, logical positivism postulates a radically new theory of knowledge - it is known as *the verifiability criterion of meaning* or *the verification principle*.

Here the inductive method is treated as an activity which gives meaning to the theoretical entities. It also implies that a word will not have meaning if there is no evidence to verify it. According to this line of thinking, logical positivists decree that the religious words like God, Soul, spirit, heaven and so on are meaningless. Since logical positivism is a combination of logical method (verification principle) and positivism (experimental evidences), it effectively links deductive propositions (Ty, H, D) and inductive propositions (T, I) resorting to process view. Such activity of giving meaning to theoretical entities on the basis of experimental evidences eliminates the dichotomy of the two classes of propositions. The TyHDTI diagram given in section 4.1 illustrates the methodology of logical positivism under process view.

Note that we have adopted the TyHDTI diagram, originally meant for logical positivism, to be the general form of scientific method using content view. In logical positivism the theoretical entities serve as mere instruments for generating experimental knowledge. But, through the content view of TyHDTI scheme, we can define the stages of *scientific method* in a static manner. [# 5][*].

Now it is easy to observe that the big bang theory and related inferences about cosmology would fall under the methodology of logical positivism. Here we treat the concepts like big bang, inflation, space and time as unobservable theoretical entities, which are given meaning on the basis of experimental data.

Having discussed about logical positivism as the correct methodology of quantum mechanics and big bang theory, we must consider certain contentious issues cropping up. There is no doubt that TyHDTI scheme outlines the practical method for discovering scientific laws. This approach will be correct and fruitful only if there is no mutual influence between theory and experimental observation. But this condition is generally satisfied only in the case of macro world made of atoms; it is traditionally the area of classical science. Regarding the study of quantum phenomena we can admit that there is the possibility of mutual dependence between theory and observation (testing). The preconceived notions and vested interests of the researcher will influence the formation of theory and the setting up of experimental methods.

This situation is expressed by the phrase*: observation is theory laden*. The experiments become less efficient because they have been designed in tune with the ideas of theory and hypothesis. Sometimes it is alleged that certain experiments are conducted just for confirming the theory in which the researcher has particular interest. The inefficiency of experimental testing due to the influence of theory is called **underdetermination**. In this situation, observation would not be able to decide the correctness of theory. In other words, theory is underdetermined by observation. It is similar to the example in which a convict escapes punishment by influencing the judge of Court. [# 6].

The problem of underdetermination does not manifest seriously in the experiments of quantum mechanics because the properties of subatomic particles and forces can be translated into observational terms. But the subsequent development of quantum physics is into the field of more elementary and fundamental phenomena such as quarks, bosons and strings, which are conceived in mathematical models. The properties of such phenomena cannot be translated into observable terms. Then the adoption of TyHDTI method is liable to produce serious case of underdetermination. We will examine this problem more closely in the next section dealing with the theories of quantum cosmology. Similarly, the latest theories under modern biology pertaining to the basic structure of life as well as those in neuroscience and psychology do not satisfy the requirements of logical positivism. Most of the experiments in these areas are theory laden and are guided by non-experimental factors linked to religious beliefs and social compulsions.

With respect to the **source** of scientific propositions, logical positivism adopts the process view of mind. Accordingly, the mental states such as ideas, emotions, freewill and memory are interrelated activities which are denoted by concerned theoretical entities. It is in contrast with the content view of mind in which mind is a nonphysical entity that exists with various component parts. The process approach of mind consists of new philosophies of mind such as *behaviorism* and *computer model functionalism*. Since these are materialist theories of mind, we can find many philosophical drawbacks, to be explained in the companion book *Life and Mind*.

Karl Popper (1902-1994) wanted to challenge logical positivism by addressing the question: when observations are theory-laden, how can a scientist objectively test the hypothesis and arrive at a valid law? In fact, underdetermination has aggravated the problem of induction and led to the collapse of verification principle. In this situation, Popper has famously proposed his *falsification theory* as a means to solve the problem of induction. He presents an alternative interpretation of scientific method: *The true nature of scientific hypothesis is that it can be refuted or falsified by a single negative instance.* Hence the falsification of hypothesis is the objective of experiment, rather than verification.

According to Karl Popper, the true criterion for forming a scientific law is falsifiability instead of verifiability. The problem of induction arises just because of the requirement that a scientific law must be verifiable by means of all future observations. Since, we are not sure about the future, there is no way of certainty about the future verification of law. In contrast, Popper argued that a scientific law holds well so long as it is not falsified by counter evidence. In this way, Popper suggests a new meaning of justification and truth with regard to scientific method. Further, Popper asserts that the ingenuity of a scientist lies in forming theory, hypothesis and law-like-statement which face the possibility of falsification by experimental evidence. So long as such conjectures have not been falsified by data, the law-like-statement is regarded as a scientific law. Popper introduces the term *corroboration* to refer to the acceptance of theories that are not falsified as yet. According to him, a corroborated theory is efficient in predicting future events by virtue of concerned laws.

Popper holds that a theory is to be considered as valid if the consequent inductive laws are not falsified. In other words, such theories can be treated as proper justification for the concerned scientific laws. Good theories will survive so long as the concerned inductive laws are not falsified. Additionally, this offers suitable criterion for selecting the better theory from among the competing theories. It is clear that Popper wants to save the rationality of science, because it is reasonable to treat a theory to be good so long as it is not refuted by experiments. The main implication of falsification principle is that it cannot be applied to propositions of metaphysics, religion and pseudo-sciences. Popper treats falsifiability as the point of *demarcation* between science and non-science. For example, propositions of religion, mythology and art cannot be falsified by objective evidences.

As a criticism, we may mention here that the falsification theory cannot be applied in a wide range of scientific investigations on account of the problem of underdetermination as defined earlier. When experiments are influenced by a particular theory, it is not practically possible to falsify the theory in a convincing manner. By suitable changes in hypothesis, the scientist can save the theory from the risk of falsification. The mutual dependence between theory and observation, as occurring in the advanced areas of science and cosmology, goes against the validity of falsification theory. In short, Popper has not succeeded in solving the traditional problem of induction. Accordingly the distinction between verification and falsification is very thin, as both are in the shadow of underdetermination.

4.3.2 Justification and Truth of Quantum Mechanics

In the context of the process view of logical positivism, there are two conflicting theories of justification – *instrumentalism* and *naïve realism*. The proponents of the former theory hold that the so called subatomic particles and basic forces are mere theoretical entities of quantum mechanics; they do not represent existing entities. However, this view is strongly challenged by the proponents of naïve realism who hold that subatomic phenomena exist really. The discussion of the concerned theories of justification is postponed to section 5.3 in next chapter. The theory of truth is reserved for the appropriate parts of the forthcoming two books.

4.4 Philosophy of Quantum Cosmology -- Special Issues

We have sufficiently explained in previous chapter the main ideas of quantum cosmology which deals with the first, second and third levels of physical world. These levels are composed of mysterious entities, quantum gravity and standard model respectively, which together is termed naively as *physical reality*. We are now trying to find the answer for the question : what is the methodology, source, justification and truth in the case of quantum cosmology?

4.4.1 Methodology of Quantum Cosmology

The three levels of quantum cosmology have separate theories as explained earlier. Then scientific knowledge about these cosmological levels is obtained through corresponding *TyHDTI schemes*. Accordingly, concerned theoretical physicists claim that each cosmological theory has experimental confirmation.

The highest level of quantum cosmology (level III) consists of the particles of Standard Model such as quarks, leptons, bosons etc. These particles are combinations of matter and energy, on account of particle-wave duality. This is why the mass of a particle is denoted by the amount of energy it contains. Applying this method in the reverse way, the basic forces are translated into *virtual particles* which are nominal quantification of energy. Another important fact is that the theory about the existence of the particles in the Standard Model is based on the *group theory* of mathematics. Using the laws of symmetry, physicist could conceive about various kinds of material particles, energy particles and their sub-particles to be included in the Standard Model, which has been firstly described in section 1.4.

The details of the experimental method called Large Hadron Collider (LHC), which is set up for discovering the elementary particles of standard model, may be repeated here briefly. In LHC experiments, particles like proton or neutron moving in opposite directions at very high speed are collided and scattered. The photos are taken of the scattering of energy. By analyzing the white spots in these photos, it is decided whether the suggested particles exist. In this manner, through experiments in various stages, scientists claim that basic particles of standard model really exist. Further, extending the method of mathematical model, theoretical physicists speculate about quantum gravity, inflation and Higgs mechanism as well as the mysterious entities like superstrings, membranes and multiverse.

It is the claim of theoretical physicists that Standard Model and prior entities are supported by experimental evidence. However, in the following paragraphs we present the main points to argue that the three levels of quantum cosmology do not represent really existing components of physical world. Further, there are important reasons to show that the experimental efforts for quantum cosmology would deviate from the method of logical positivism. [# 7][*].

Point 1: *The verification principle of logical positivism does not hold good in the case of Standard Model.*

The supporting arguments are given below:

a) We know that the particles like proton and electron suggested by quantum mechanics (level IV) satisfy verification principle. The evidences for such particles are available from the macro world, i.e. level V. *But the basic particles belonging to the standard model of level III -- quarks, bosons and many other elementary particles -- are pure mathematical models, which do not follow the verification principle.* Such theoretical entities cannot be translated into phenomena appearing in the visible macro world.

b) Since the verification principle is not followed, there is a possibility to interpret the signs seen in the photos according to the whims of the scientific community, since the discovery of a basic particle would bring them wealth and fame. The rule that theory and experiment should be independent of each other is not observed here. Therefore the problem of *underdetermination* of experiments is conspicuously expressed. In this situation, the quality of experimental process becomes very low since hypothesis, deductions, experiments and observations are interrelated.

c) Because of these reasons, the principle of logical positivism – the view that scientific method is the process of giving meaning to theoretical terms – cannot be applied in the case of standard model. It is in this circumstance that the word *model* replaced the term *theory* in scientific parlance.

Point 2: *We do not have sound reason to state that standard model particles exist as elementary components of physical world in the strict sense.*

The importance of this point can be illustrated with reference to the discovery of quarks. According to standard model, there are six types of quarks. They are represented as U, D, C, S, T and B. These letters denote respectively the words Up, Down, Charm, Strange, Top and Bottom. Proton is UUD. At the same time, neutron is DDU. Is there any meaning in saying that U, D quarks are the factors of proton and neutron?

To illustrate this issue, we can take two objects water and glass. Through experiments we know that the basic factors of water are oxygen and hydrogen. Glass breaks when it falls on ground and it gets scattered. These pieces of glass are the factors (parts) of the original glass – it is said in a different sense. Here, we do not intend the chemical structure of glass. Similarly U, U, D quarks are the basic particles obtained by the way of collision and scattering of protons using the particle-wave principle. Though they are parts of protons, they are not basic factors as in the case of water. John Gribbin, in his book *The Universe: A Biography*, has written: "These are not particles that were present in any sense 'inside' the original particles and were knocked out by the collision; they are new particles that have literally been made out of pure energy" [# 8]. This means that though U and D are different packets of energy, they cannot be construed as the basic constituents of proton or neutron, which itself is a bigger packet of energy.

We shall consider once again that each of the basic particles of standard model is an energy field. As explained earlier, here field is much different from what we see in the visual world; field is a particular mathematical function used to denote features of a basic particle. This is a figurative usage, an abstract idea. This does not depend on time and space. It is in this sense that the term 'model' is used. Accordingly, we understand that U, D quarks are models only, not worldly objects.

It can be remarked that the virtual particles representing the four basic forces is a clever idea to eliminate the ambiguities of particle-wave duality. In this way, the standard model consists of particles only where each particle is a packet of energy. However, the concept of virtual particle is imaginary in tune with the method of quantum field theory; it does not satisfy the verification principle. This leads us to the conclusion that the particles of standard model cannot explain in a proper way the subatomic particles and forces included in the level of quantum mechanics.

The search for fundamental particles of third level – standard model – is the result of the obsession of theoretical physicists to find the stuff or building material of physical world. Since atoms are extended particles, it is a way of conventional thinking that the ultimate constituents of atoms also are particles. *Here the scientists have failed to learn the philosophical issues of particle-wave duality.* The LHC experiments and allied projects like neutrino research are misconceived programs that serve the self-interest motives of the concerned scientists.

Point 3: *The mathematical model of mysterious entities is a form of scientific fiction.*

The situation worsens when cosmology enters the stage of mysterious entities like superstrings, membranes and multiverse. The proliferation of such models reflects the advancement of pluralism, which is against the quest for unification underlying physical reality. *The status of mystical entities is further away from the verification principle. Such models can be aptly described as* **scientific fiction**. It is necessary for us to enquire about the arguments adopted by physicists to promote their speculative research.

Modern Phenomenalism

Since the mathematical models of quantum cosmology are speculative ideas without proper experimental evidence, can we say that such models represent invisible basic entities? It is necessary to address this issue because we normally expect that the cosmological entities must necessarily exist in order to explain the historical development of physical world. In this situation, a new methodology has been formulated to interpret the experiments and observations for discovering the basic particles. We term this principle as *Modern Phenomenalism*, which is abbreviated here as *Mophism*. [# 9][*]. It is the argument that mathematical functions denote the phenomena of quantum cosmology.

The roots of this line of thought can be found in the classical doctrine of phenomenalism famously advocated by George Berkeley (1685-1753). Formally we can define phenomenalism as the doctrine that human knowledge is confined to the appearances (phenomena) presented to the senses.

George Berkeley, as a proponent of empiricism, argued that we get knowledge about the features of objects only through our senses. That is, we know about the objects by forming ideas about their qualities (length, breadth, weight, colour etc). But we cannot say that there is an internal material object which holds these qualities because there are no sensory experiences about it. Even if we divide the object repeatedly, we can experience the sensory qualities only. According to this view, an object is the sum total of its qualities perceived by mind. In other words, all these qualities are ideas originated in the observer's mind. Berkeley upholds the subjective characteristics of knowledge by arguing that all objects in the physical world are mere ideas in our mind. Material object is a phenomenon (the totality of qualities experienced). The scientific knowledge about matter is empirical; the question whether matter really exists should be avoided from science. To exist means to be perceived (*ese est percipi*) – this is Berkeley's motto.

We shall propose that mophism, the new version of classical phenomenalism, is the methodology adopted to study the basic levels like standard model, quantum gravity, superstring, membrane, multiverse etc. The mysterious phenomena have ten or more dimensions and these are completely mathematical. They are beyond the limits of scientific experiments, which can deal with objects existing in four dimensions of space-time only. Theoretical physicists in recent decades are designing highly expensive experiments for proving that extra-dimensional phenomena exist concretely. *They choose particular hypothesis in artificial manner; such hypothesis is not the observational translation of the mathematical models or theoretical entities. Moreover, the evidences are collected in an indirect way, without the support of verification principle; so they are not trustworthy.*

As an example, experiments on dark energy and dark matter may be considered. How can we say for sure that there are experimental evidences on these phenomena which are beyond the realm of our physical world? It is similar to the quest to find souls and angels in this world. There is no sound basis for mophism as the methodology for the experiments on quantum gravity and other mysterious entities.

We may recall that George Berkeley and David Hume put forward empiricism based on content view about macro things of physical world. But postmodern models of quantum cosmology are different types of mathematical functions pertaining to the invisible world. Mophism considers them as phenomenally existing objects. In this form of radical empiricism, the problem of **skepticism** grows deeper. When physicists say, with the help of experiments conducted indirectly, that the fundamental objects exist phenomenally, it is difficult to overcome the objections of skepticism. Further note that there is a contradiction in the usage *physical reality* which implies that basic phenomena of science are real. This terminology reflects the vested interests of scientists as well as their dislike for philosophy. [# 10] [*]

The above said defects are very serious. It is to be reiterated that the field of science is the physical aspect of universe, without considering its mental or nonphysical part. Hence, mophism follows the agenda of theoretical physicists for explaining the origin of our universe using the method of materialism and empiricism. Since physical world is an everyday experience, scientists are inclined to assert that the original cause of universe is physical reality consisting of standard model and mysterious entities.

Consequently, the issue of reconciling the diverse propositions under DP and IP would come up in the case of quantum cosmology also. The methodological dilemma in quantum cosmology is similar to that of classical science because the tenets of Modern Phenomenalism (Mophism) can be challenged by the doctrine of rationalism. Next, let us turn to the topic of *source* pertaining to the models of quantum cosmology.

4.4.2 Source : Materialist Philosophy of Mind

The most modern development of psychology is cognitive science, which assumes a materialist theory of mind. In this context, the doctrine about the origin of mental states is called **computer model functionalism;** its key points are given as following. Note that a computer works in the desired way as per a program; so it is an interplay between hardware and software. Accordingly the computer *appears to work purposively* knowing the meaning of words or symbols given in the program or software. In an analogous manner, mind is treated as a computer; the brain and other parts of nervous system (BNS) together are treated like hardware, which produces diverse neuron networks. *The algorithms of such networks are interpreted as different mental states*, which are observed through various types of behavior. This is a physical, practical and interpretative approach for describing the production of mental states; we can call it as the computer model of mind.

The method of conceiving algorithms as existing entities is in accordance with *Modern Phenomenalism (Mophism)* described above. We may reconsider the question about the origin of mental algorithm. Since computer model functionalism is the process version of materialism, we can say that it moves towards *epiphenomenalism* in order to explain the formation of various mental states. The suggestion is that the mental states or algorithms are byproducts of the physical process in brain and other parts of nervous system. There are certain defects in this view: If mental states are the byproducts of physical processes, the concerned algorithm is mechanical. It cannot account for the mental causation and creativity that produces the motivation to do purposive actions. In other words, the nonphysical consciousness is outside the ambit of physical algorithm.

To complete our discussion about the epistemology of quantum cosmology, we must consider the issues of justification and truth also. As per our earlier decision, we will present it in due course.

4.5 The Crisis in Philosophy of Physical Science

As the final part of this chapter, we may recapitulate the key issues with regard to the scientific knowledge about physical world.

The ontological doctrines of Deism, Materialism (Naturalism) and Physical Process View are based on divergent worldviews suggesting different conceptions about the existence of physical world. The dual existence of matter and mind is an intractable problem in the streams of philosophy pertaining to classical science. More importantly, the process approach of quantum physics adopts only an interpretative method focusing on the activities of physical entities; so it fails in addressing the ontological question. To tide over the philosophical dilemmas, we will have to continue our search for a unifying theory of reality.

From our preceding analysis, we find that *philosophy of science* can achieve its objectives only if the conflicts and dilemmas shown in the table below are overcome.

Specifically, we require answers to the weighty questions: Is there reliability in the contingent truth available from scientific method? How can we explain the progress of science achieved so far? However the problems of induction and justification would persist as the bones of contention to be tackled with greater intellectual effort. Our systematic presentation points to the need of recasting philosophy of science from content view, in order to remove the inconsistencies in ontology and theory of knowledge pertaining to different branches of science. We can hope that the reconciliation of the opposite doctrines will be achieved through the innovative System Philosophy of Science.

I would adopt the convention that the word 'philosophy' refers to the monistic philosophy developed by other philosophers in the bygone centuries. The new approach of philosophical analysis introduced in this book is specially designated as System Philosophy.[*].

Table : Main Doctrines and Conflicts [# 11][*]

	Conflict between Rational and Empirical Doctrines under Content View	Conflict between Rational and Empirical Doctrines Under Process View
Ontology	Deism versus Materialism and Naturalism	Rational Pantheism v. Physical Process View and Mysticism
Methodology	Rationalism versus Empiricism	Rationalism versus Logical Positivism
Source	Idealism versus Epiphenomenalism	Idealism versus Behaviorism and Computer Model Functionalism
Justification	Metaphysical Realism versus Naïve Realism	Metaphysical Realism versus Scientific Realism
Truth	Necessary Truth versus Contingent Truth	Necessary Truth versus Pragmatism

Religion, Ethics and Scientism

A major issue to be tackled by our philosophical quest is the synthesis between science and religion. Can existence of matter be accepted along with the notion of religious God? This question often leads to the hostility between atheists and believers. Similarly there is considerable difference in the language and description of cosmology advanced by scientists when we compare it to the version of theologians. We desire that the conflict between scientific knowledge and religious belief must be settled coherently.

There is a common refrain that scientific theories and laws are factual propositions, and the considerations of ethics are outside the realm of scientific thought. It is definitely a misconceived and parochial idea, since we have established that theory and experiment are influenced by political, religious and other social forces. The purpose of scientific research must be evaluated applying the norms of ethics. Science is not value free. The impact on environment and social justice deserve the judicious attention of all involved in the scientific project.

It may be remarked that the hitherto books on philosophy of science lacks clarity in their contents; hence the above table would be greatly helpful to the readers. The avowed aim of the foregoing discussion has been to investigate into the defining features of physical science and its limitations. We are interested to compare and contrast between physical science and certain forms of non-experimental knowledge such as astrology, palmistry, parapsychology and holistic medicine which are generally dismissed by scientists as *pseudo sciences*. In this context, it is important to consider the ideological position called **scientism** that is a professional attitude of many prominent scientists. Alex Rosenberg has described scientism as : *The unwarranted overconfidence in the established methods of science to deal with all questions, and the tendency to displace "other ways of knowing" even in domains where conventional scientific approaches are inappropriate, unavailing or destructive of other goals, values and insights.* [# 12].

It must be admitted that through scientism the proponents want to uphold the cognitive superiority of science in comparison to other forms of knowledge. The severe conflicts in the epistemological doctrines discussed so far indicate clearly the fallacy of scientism. We can remark that the huge expenses incurred in the experiments of theoretical physics – Large Hadron Collider, neutrino research and so on - are clear examples of the self-interest motives of scientists because there are doubts about the explanatory power of such knowledge.

To end the present chapter, we notice that there is a multiplicity of outstanding issues in the field of philosophy of science. Accordingly, the discipline cannot be regarded as a work in progress; rather it is a mélange of confused doctrines. The search for the synthesis of philosophical conflicts will take us to higher levels of deliberation in the ensuing chapter.

NOTES of Chapter 4

#1 The main reference books used here are Bird (2003), Grayling (Editor) (1995), Hospers (1997), Lavin (1989), Rosenberg (2000), Tarnas (1991) and Urmson J.O. and Jonathan Ree (1989). Note that this chapter innovatively recasts the philosophy of science under the topics of *methodology, source, justification* and *truth*.

#2 The clear cut division into mechanistic and process philosophies of science is an original idea introduced in this chapter. The illustrious books of reference lack a proper description of the main branches of philosophy of science. As the best example of the confusion in this regard, see Rosenberg (2000), pages 1-18.

#3 Here, we adopt a practical meaning of *a priori*, which defers from the definition given by rationalist philosophers, notably Descartes and Immanuel Kant. Descartes dealt with deductive propositions; he held that *a priori* refers to the propositions known on the basis of reason alone and without any sensory experience of the world. Immanuel Kant distinguished between *analytic a priori* and *synthetic a priori*. See Hospers (1997), page 236.

#4 The phrase *TyHDTI scheme* is my original idea. The discussion about scientific explanation is excluded here due to space constraints. In due course, we will realize that *TyHDTI scheme is the common method for all kinds of knowledge.*

#5 The conversion of the process view of logical positivism into the content view of TyHDTI scheme is my original idea. This is the central point in the epistemology of System Philosophy.

6 Underdetermination is alternatively and more famously known as *Duhem-Quine Thesis*, named after the exponents Pierre Duhem and W.V.O. Quine. This thesis exerts serious challenge to the verification principle of logical positivism.

7 These critical points are my original ideas.

8 Gribbin, John (2008), *The Universe: A biography* (Penguin Books, London), page 12.

9 See Urmson J.O. and Jonathan Ree (1989) for the definition of phenomenalism. Here, we consider only the version advanced by George Berkeley, even though David Hume, J. S. Mill and Bertrand Russell have attempted to explain this doctrine in their own empirical ways. The innovation proposed here is the notion of *modern phenomenalism* (*this is my original idea).

#10 The popular notion of *physical reality* stands for the most fundamental aspects of physical world, especially the standard model particles and mysterious entities.

11. This table would serve as the impetus for developing System Philosophy in subsequent chapters.

12. See Rosenberg (2000), pages 7. Further, this book gives a comprehensive criticism of scientism in pages 7, 164-165 and 173. The ideology of scientism holds that true knowledge about world comes through science only. It will be explained in next chapter that the argument of scientism follows from scientific realism.

Chapter 5

System Philosophy of Science

****

5.1 Summary of Dilemmas in Methodology and Source

5.2 System Philosophy about Methodology and Source

5.3 Overview of The Problem of Justification in Physical Science

    5.3.1 Classical Science

    5.3.2 Quantum Mechanics

    5.3.3 Quantum Cosmology

5.4 System Model of Justification

    (Existence of Physical World)

5.5 Spiritual Science of Ancient and Medieval Periods

Author's main original ideas are marked by [*].

The mark [#] gives the number of note at the end.

****

As already introduced, *philosophy of science* is the epistemology or theory of knowledge pertaining to scientific propositions. The main components of this subject are methodology, source, justification and truth. We are going to deal with the propositions about fundamental aspects of physical world; hence its epistemology would serve as the basis for developing the integrative philosophy of all levels of science as a whole.

In this ground-breaking chapter, we will try to reconcile the conflicting doctrines of methodology, source and justification with regard to physical science. Since the topic of truth requires a comprehensive treatment, taking into account other subjects also, it is postponed to a later occasion. [# 1].

Recognizing the incomplete state of philosophy of science, we would propose the principles of System Philosophy in order to derive the innovative aspects of scientific justification. The key achievement is the system model of phenomenal existence of matter and physical world. These tenets can be used for completing the epistemology of all levels of physical science. This will lead to the solution of the cosmological puzzles of matter, big bang, space, time, superstrings, membranes, dark matter and dark energy, which is reserved for next chapter. The important highlight of our thesis is that we can explain the origin and evolution of physical world without resorting to the intelligent design argument. In contrast, the final part of this chapter presents a brief review of spiritual knowledge about world, which prevailed in ancient and medieval periods.

5.1 Summary of Dilemmas in Methodology and Source

Technically speaking, the laws about cause-effect relations and other sensible properties of physical things constitute our scientific knowledge. It is helpful to recall a few key points to familiarize with our new approach of analyzing various scientific concepts and propositions.

The scientific laws are generally obtained through the sequence of theory (Ty), hypothesis (H), deduction (D), testing (T), and inductive inference (I). These stages are ordered like the organs of an animal. The propositions coming under successive stages of Ty, H and D together is called *deductive propositions (DP)*; the propositions of T and I are collectively designated as *inductive propositions (IP)*. We have suggested the new phrase *TyHDTI scheme* to denote the scientific method of combining DP and IP. In this context we can list the connotations of the four components of epistemology as following:

- ***Methodology*** is the deliberation about the general components – *Ty, H, D, T and I* -- of scientific method as well as about the relative importance of DP and IP in the meaning of scientific laws. Also, the various definitions and meanings given to the fundamental terms are clarified in different theoretical situations.
- ***Source*** denotes the theory about the structure of scientific mind, which generates the diverse kind of propositions under DP and IP.
- ***Justification*** deals with the issue whether the scientific law represents actually existing aspects of universe. Accordingly, we must get sufficient evidences to judge that the concerned scientific law is valid. The essence of justification is that the components of theory (Ty) -- namely space, time, matter, energy, and so on -- as a whole represent the reality of physical world. It finally leads to the question: does matter or physical world exist?
- ***Truth*** is the quality of a justified belief under the TyHDTI scheme, when it conforms to the actually existing things of universe. Moreover, truth is the unifying principle applicable to all kinds of knowledge such as science, religion and art.

We know that **classical science** adopts content view; its theories are based on the principles of mechanistic worldview. Consequently, its propositions under DP and IP are not unified, since these are treated to have separate kinds of validly and meaning. Then there are two opposite doctrines – rationalism and empiricism -- about **methodology** of the physical laws.

Through the detailed analysis of rationalism and empiricism, conducted earlier, we can state that these opposite positions cannot be reconciled within the framework of mechanistic worldview. According to rationalists, various theories of classical science are the result of abstract thought, without the aid of empirical facts. This view can be objected by pointing out that such theories may not be proved by experimental data. For example, Isaac Newton proposed that the origin of gravitational force is God; this theory does not have scientific validity. Similarly classical physics has no experimental evidence to support the principle that pure matter (atoms) exists. On the other side, empiricism also is a failure due to skepticism and problem of induction. The empirical theories about the fundamental properties of matter suffer from lack of explanatory power.

Next to be considered is the methodology and source of **quantum mechanics**, which deal with the laws about four subatomic particles and four basic forces. Since these subatomic phenomena have the wonderful property of particle-wave duality, the names of subatomic particles and forces do not have representational property. For example, we cannot say that the word *electron* represents a specific particle. In this situation the mechanistic worldview is not applicable. Alternatively, scientists and philosophers of quantum mechanics shifted to physical process view; they proposed the methodological doctrine called *logical positivism*.

Rather than assuming the existence of particular subatomic particles and forces, quantum mechanics describes the activities of subatomic phenomena and formulates the cause-effect laws. The words denoting the subatomic phenomena are treated as *theoretical entities*. The activities of these phenomena can be observed by showing the effects in the atomic world or macro level. For example, the emission of electrons is inferred from the working of the picture tube of TV. Such indirect evidences would give meaning to the theoretical entity; this will serve the pragmatic objectives of subatomic laws. In this manner, logical positivism postulates a radically new theory of knowledge – it is known as *the verifiability criterion of meaning* or *the verification principle*.

The methodology of logical positivism, resorting to process view, effectively links deductive propositions (Ty, H, D) and inductive propositions (T, I) in the form of an activity of giving meaning to theoretical entities on the basis of experimental evidences. We have introduced the phrase *TyHDTI scheme* to denote the linking process -- thus the dichotomy of the two classes of propositions DP and IP is eliminated.

With respect to the **source** of scientific propositions, logical positivism adopts the process view of mind called *behaviorism*. Accordingly, the mental states such as ideas and emotions are treated as theoretical entities, which are given meaning through outward behavior. It is in contrast with the content view of mind in which mind is a nonphysical entity that exists with various component parts. The behavioral approach to study mind suffers from many philosophical drawbacks; it will be explained in the companion book *Life and Mind*.

Due to the above issues of logical positivism and behaviorism, philosophers want to shift from process view to content view regarding methodology and source of quantum mechanics. Then we are back to the field of rationalism and empiricism as in classical science. We are inclined to treat electron, proton, etc as existing entities with particle-wave duality in subatomic world. Consequently, the issue of reconciling the diverse propositions under DP and IP would come up in the case of quantum mechanics also.

Coming to the stage of **quantum cosmology**, its methodology is Modern Phenomenalism (Mophism) as explained in previous chapter. It follows the agenda of theoretical physicists for explaining the origin of our universe using the method of materialism and empiricism. *Consequently, the issue of reconciling the diverse propositions under DP and IP would come up in the case of quantum cosmology also. The methodological dilemma in this stage is similar to that of classical science because the tenets of Mophism can be challenged by the doctrine of rationalism.*

Next, let us turn to the topic of *source* pertaining to the models of quantum cosmology. Here the philosophical problem is related to the most modern development of psychology called cognitive science, which assumes a materialist theory of mind. In this context, the scientific doctrine about the origin of mental states is *computer model functionalism*. Anticipating the discussion about philosophy of mind in the companion book, it is indicated here that the conflict between rationalism and empiricism must be resolved in the study of mind also.

We have briefly reiterated the key points about the dilemmas regarding methodology and source of *classical science, quantum mechanics* and *quantum cosmology* in this order. System Philosophy can reconcile these conflicts; this is the topic of the ensuing sections. An overview of the problem of justification will be given in section 5.3; it is the stepping stone to formulate the *system model of justification* in the subsequent section.

5.2 System Philosophy about Methodology and Source

We may start this discussion by considering the **three rules of thought** originally proposed by Aristotle (BC 384-322). These are given as below.

1. Rule of Identity : A is A
2. Rule of non-contradiction : A is not both B and not-B
3. Rule of excluded middle : A either is B or is not-B
   (a proposition is either true or false)

Aristotle holds that everybody must apply these laws of thought for knowing the individual things separately. It means that we can think about things – particles or metaphysical beings – distinctly as parts of universe. Moreover, we assume in a common sense way that such things have existence. In this manner, for example, we hold that the words like paper, pen, cow, cat, soul, heaven, hell and God represent distinct objects. As a corollary, the opposite entities are assumed to exist as separate objects. So it is assumed that day and night (or man and woman) are opposite objects which have separate existence.

The fundamental tenet of System Philosophy is the definition of **system** as a productive structure composed of opposite entities. The opposite components of a system are similar to the X axis and Y axis pertaining to the coordinate model of analytical geometry. This system model is illustrated by a diagram to be shown in the second chapter of next book *Life and Mind*. Hence, system is a symmetrical structure of X-Y coordinates. Here X and Y have dialectical and productive relation similar to that of a factory. *As a rejection of Aristotle's rules of thought,* the opposite entities X and Y are not independent entities. Instead, X and Y are opposites having complementary character. We cannot define one entity without considering its opposite entity. The opposite components of a system are complementary to each other; they have interdependent existence. [# 2][*]

As explained in the third chapter, we can divide the physical world into five levels – three levels of invisible world and two levels of visible world. On account of this state of affairs, there are five main levels of theories for physical science to produce knowledge through TyHDTI schemes – these are conveniently classified into quantum cosmology, quantum mechanics and classical science. In this context, the philosophical problem of methodology and source boils down to the requirement that we must effectively unify the propositions under deductive propositions (Ty, H, D) and inductive propositions (T, I). Such a synthesis would come from the *system philosophy of mind* to be developed in the companion book. In anticipation of that doctrine, the relevant points are presented in the following paragraphs.

The propositions under deductive propositions (DP) are traditionally treated as *rational* because it involves the abstract concepts produced by the creativity of human mind. On the other hand, inductive propositions (IP) consist of experimental observations and inferences; these are traditionally regarded as forms of *empirical* knowledge. The absolute division of scientific knowledge into such two classes of propositions resulted in the conflict between rationalism and empiricism.

In the context of system philosophy of mind, we propose that human mind is a system formed by the dual aspects of *brain and other parts of nervous system (BNS)* and *consciousness*. Mind has many levels which are mainly classified into scientific mind, religious mind, artistic mind and philosophical mind. Each level of mind corresponds to a part of BNS and consciousness. At present we are focusing on scientific mind which is a system with X-Y model, where X and Y denote the concerned parts of BNS and consciousness respectively. So we can conceive scientific mind like a factory for producing two classes of propositions denoted as DP and IP.

Now we know that DP and IP are two levels of propositions where both have the aspects of X and Y as indicated above. The proportion between rational aspect and empirical aspect is higher for DP as compared to IP. The process of generating DP consists of more rational thinking and less use of empirical data. On the other hand, IP has less measure of rational part and more amount of empirical part. *Though DP and IP can be distinguished as two levels, they are interrelated for producing a system of meaning.* This principle is elaborated in a different way using the notion of "holism of meaning". In this situation, DP and IP are complementary opposites which constitute a whole of knowledge as per the TyHDTI scheme. [# 3].

Further it may be added that the traditional dichotomy between the notions of rational mind and empirical mind is a fallacy caused by the application of Aristotle's rules of thought. The correct view is to hold that the so called rational mind and empirical mind are two levels in the system of BNS and consciousness. As an extension of this line of thinking we can reconcile the opposition between necessary truth of DP and contingent truth of IP.

The above innovative ideas would enable us to make a fresh assessment of the **problem of induction,** which was mentioned in previous chapter. It may be pointed out that logical positivism adopting process view was an attempt to solve this problem. In section 4.1 we mentioned that Carl Hempel described scientific method as Deductive-Nomological Model (D-N Model), which is alternatively called as Hypothetico-Deductive method. This model effectively links the two classes of propositions, namely deductive propositions (Ty, H, D) and inductive propositions (T, I). We can note that, Hempel's interpretation is valid only through the verifiability criterion of meaning. But, when we consider the TyHDTI scheme under content view, the problem of induction persists. The attempt of scientists to produce a universal law by using theory and a set of experiments cannot overcome the skeptical arguments of David Hume.

Of late, some empiricist philosophers have proposed the argument called *reliabilism*. Accordingly, they take into account the everyday meaning of the term *rational*. For example, if we have sufficient data about the past occurrence of the weather condition of two consecutive days including yesterday and today and if we adopt logical reasoning, then there is a possibility of forecasting the weather of tomorrow. This procedure is treated as rational, as per the ordinary usage of the term. Proceeding in this line of argument, the proponents of reliabilism holds that inductive method is reliable and hence rational. However, we can suggest that this artificial method of circumventing the problem of induction is not satisfactory.

Let us take a close look on the problem of induction advanced by David Hume, who was mainly concerned with the inductive method involving cause-effect relations. Since he did not recognize the creativity of human mind, he failed to give due importance to *theory* in the process of formulating scientific law. This can be interpreted now as the principal reason for his skepticism. **Our idea of solving the problem of induction, at least up to a comfortable extent, is presented below.**

If we have a good theory, it is possible to feel that there is a *necessary connection* between cause and effect. Consider the example of a tree climber who falls freely from the height of thirty feet. This fall would naturally cause the breaking of his bones. Here, the cause is the fall from the height of thirty feet, while the effect is breaking of bones. Now we can see a sort of necessary connection as cause-effect, on the basis of a good theory about the anatomical structure of human body. It is possible to hold that, in this case, the problem of induction is partly removed by using a **good theory** as per the TyHDTI scheme. In other words, a good theory would save a scientific law from the problem of induction up to a reasonable extent. However, there is still an element of uncertainty about the repetition of scientific law in future instances because future is unknowable per se. Considering the role of a good theory, we can suggest that there is not much significance to Hume's problem of induction.[*].

5.3 Overview of Problem of Justification in Physical Science

For a comprehensive treatment of the justification problem in physical science, this section presents a quick survey of the concerned issues with respect to the three paradigms - classical science, quantum mechanics and quantum cosmology. The final solution to the question whether physical world exists actually is reserved for the subsequent section.

5.3.1 Classical Science

We may recall that *justification* means the validation of a law on the basis of sufficient evidences. It involves the production of arguments to prove that the state of affairs represented by the law exists actually. For example, the scientific law about rain is justified only if we can hold that rain exists as a part of this universe; this will finally rest on the proposition that matter exists. Classical science was developed in 18th and 19th centuries on the basis of mechanistic worldview. Its most basic principle, as given in section 1.2, is that the physical world exists as a giant machine made of atoms. All atoms consist of homogeneous substance and it is denoted by the word *matter*. The essential property of matter is extension or mass.

As per present knowledge, there are about 110 kinds of atoms differing in size. Now we may consider the visible and invisible levels of physical world, illustrated by Table 1 of chapter 3. Then the atoms, molecules and higher substances constitute the macro world or Level

V. For making the structure of atom or other substances and for its motion, energy is required. In fact, atom is the union of two opposite components namely matter and energy. Taking this fact into the context of Classical Science, the problem of justification can be expressed by the questions: Do atoms exist really? And why are there different kinds of atoms that are combinations of matter and energy?

Scientists hold that atoms have extremely small size, to the order of one by millionth of a centimeter for diameter. Hence, atom is practically unobservable through our sense organs. In this situation, we must consider the opposite doctrines of methodology -- rationalism and empiricism – with regard to the propositions about atoms. Correspondingly there are two conflicting theories of justification, namely **metaphysical realism** and **naïve realism.**

As per the metaphysical theory, the existence of atoms can be deduced from the axioms and definitions pertaining to mechanistic worldview. The concerned specific arguments belong to the philosophy of Descartes. But Immanuel Kant(1724-1804) devastated the metaphysical realism through his famous statement: *existence is not a predicate*. The implication of Kant's philosophy is that the theoretical concepts about atom belong to the realm of phenomenal knowledge; it does not stand for reality. **A clarification** may be given about this key idea. When we say that atom is our mental construction, it should not be construed as an imaginary object like angel or ghost. On the other hand, there is an unknown reality for physical world, which is transformed into the entity called atom for practical purposes. So we understand the phenomenal aspects of physical world using the notion of atom.

It is pertinent to state the additional drawbacks of metaphysical realism in the context of Classical Science as following.

> ➢ If various kinds of atoms exist separately and eternally, it calls for an explanation. Also we must take into account the *mind-body dualism* at least in the case of human beings. Rationalists tried to convert the theological position called deism into an ontological theory; but it has many inconsistencies from philosophical perspective.

> The separate existence of matter and energy has been refuted in 20th century through the development of modern physics. As such, the mechanistic worldview is not tenable now.

The next step is to examine the theory of justification related to the methodology of empiricism. Here the ontological theory is materialism or naturalism. It is helpful to recollect the essential ideas pertaining to these doctrines as described in the companion volume *Discovery of Reality*. David Hume asserted that scientific laws, in the form of cause-effect relations, are derived from our psychological habit of association of ideas. He provides legitimacy to the knowledge obtained through induction. And, the deductive propositions like definitions and other abstract concepts are treated as mere abbreviations about sense data. In this situation, the question whether matter and physical world are real cannot be answered affirmatively. We can perceive a thing only on account of its physical properties observable by our sense organs. There is no sensory experience of the substratum called matter.

Then the idea of matter is only an abbreviation of such qualities as length, breadth, volume and weight, which we experience in physical things. So there is no justification for holding that matter exists really. Science is a framework of physical laws pertaining to phenomenal qualities of world and it is patently incapable of talking about reality. As a consequence, the realist position of materialism or naturalism cannot be maintained. In this situation, David Hume falls into *skepticism* about the real existence of physical world.

However, the exponents of classical science continue to maintain that the physical world exists really. They argue that matter exists fundamentally in the form of atoms because the laws of classical science are highly successful in explaining the visible phenomena such as heat, sound and motion of astronomical bodies as well as the chemical reactions of various substances. Accordingly, the theory of justification under classical science is called **naïve realism**. It is a practical assumption adopted by scientists ignoring the skeptical arguments of David Hume. We can agree that *naïve realism* is a problem in philosophy of science;

scientists choose to continue with their empirical studies without considering its philosophical drawbacks.

5.3.2 Quantum Mechanics

This level of physical science aims to describe the activities of invisible entities such as four subatomic particles and four basic forces, which together accounts for the visible phenomena of physical world. How can physicists justify the laws of quantum mechanics since they deal with the invisible aspects of subatomic world? This question has deep ramification and we will discuss about it concisely.

Since the activities of subatomic particles and basic forces can be translated to the effects in physical world, we may assert that such micro objects must have existence. Note that this talk about existence belongs to the area of empiricism coming under content view. Obviously it creates tension with the process view of logical positivism. In this situation, there are two conflicting theories of justification – *instrumentalism* and *naïve realism*. The proponents of the former theory hold that the so called subatomic particles and basic forces are mere theoretical entities of quantum mechanics. They do not represent existing entities; instead they must be treated as instruments for supporting the evidences from physical world in accordance with the verification principle. The purpose of such scientific laws is the description of the activities of physical things, without explaining the ontological aspect. This position is a version of *pragmatism* with regard to scientific knowledge and it is alternatively termed as *antirealism*. A group of illustrious quantum physicists, mainly, Werner Heisenberg, Neils Bohr and A. N. Whitehead upheld instrumentalism. However, they were strongly challenged by the proponents of naïve realism with Albert Einstein in the forefront.

As a matter of fact, empiricism is the theory of knowledge under content view about the visible world made of subatomic particles, forces, atoms and molecules. The scientists holding naïve realism want to extend empiricism to the subatomic world of *quantum mechanics*. The naïve realism achieved dominance over instrumentalism because it provided strong encouragement to the experimental designs of theoretical physics as developed in the recent six decades. But the problem of induction and skepticism would challenge this doctrine seriously; it is a legacy of empiricism. The serious drawbacks of naïve realism as a theory of justification can be easily identified.

In this context, we have to reflect further upon the question whether the structure of matter (atoms) is ultimately in the form of subatomic particles and forces. Since naïve realism is only a pragmatic assumption aligned with empiricism, the issue of justification cannot be solved at this level. Here we have to consider the recent discoveries of theoretical physics about more elementary and invisible entities such as quarks, bosons, superstrings, membranes, dark matter, dark energy etc. Thus the structure of physical world is now analyzed into deeper levels. These are the models of *quantum cosmology* containing more complex epistemological issues of scientific realism; we postpone this subject to next section.

Process philosophers cannot unify the levels of theories pertaining to different branches of science dealing with inanimate and biological world. (Note that the recognition of levels and the issue of unification belong to content view of knowledge). We must admit that the pluralism of theories and corresponding methodologies cannot be explained under process view. In this predicament, the concerned process philosophers are restrained to hold that *there is no scientific method*. [# 4].

Faced with various issues outlined above, the process philosophers adopting *instrumentalism* are unable to explain the ***progress of science*** achieved in recent centuries. To overcome this crisis situation, Thomas Kuhn famously suggested that progress of science maybe accounted from the evolutionary and social perspective. In his path-breaking book *The Structure of Scientific Revolutions* published in 1962, Kuhn analyses the historical progress of science. This book has been very influential in the recent period because it highlighted the following principles:

a) Different levels of scientific methodology are interpretative structures called paradigms.

b) A paradigm can undergo change *historically* due to social and psychological factors.
c) The progress of science is due to the emergence of stronger paradigms, rather than the preciseness of scientific method.

Adopting evolutionary perspective, Kuhn describes the origin of a new paradigm replacing the old one, which was found unsuitable due to the overwhelming presence of anomalies. Such emergence of new paradigm is treated as a revolution in the history of science. Kuhn's historical view of paradigms found application in the study of change related to a wide range of academic disciplines – the resulting doctrines collectively advance the movement called *postmodernism*.

It can be admitted that the historical approach has certain advantage as it enables us to study the social-cultural-political dimensions of scientific enterprise. For example, the development of telephone in the course of last century was mainly due to the exigencies of war and business. Also the psychological effects of telephone on people of different age groups can be analyzed within the same framework. In contrast, deliberation about scientific method runs into difficulties on account of the conflict between naïve realism and instrumentalism, as explained earlier. In this situation Kuhn and followers have taken an extreme position to hold that history of science is the proper part of philosophy of science, there by relegating the role of methodology and related aspects of traditional epistemology.

Now it is pertinent to state the important points of criticism against the doctrine of Thomas Kuhn. The historical view essentially promotes the *relativism* about methodology and truth, causing great confusion of thought. It aggravates the problem of pluralism about theory and methodology. Further the approach of historical process cannot recognize the levels of theories pertaining to different branches of science. Clearly, it is an aberration from the unifying approach of philosophic enterprise. [#5].

To conclude this part, I may add a few points: Mechanistic Worldview can be extended to subatomic phenomena, by defining the elementary forms of particles and energy. Then empiricism is the methodology of quantum physics. But it leads to skepticism. So we must adopt system philosophy of science being proposed in my book. Kuhn's historical view cannot talk about existence of things. Moreover he does not answer the question: what is the method of science? In this situation, we have to treat both mechanistic view and historical view as complementary streams of philosophy of science. One approach cannot replace the other.

5.3.3 Quantum Cosmology

For justifying scientific propositions, under content view, generally we require that the qualities or features of physical things must exist upon a pure substance called matter. The concerned ontological theory is materialism or naturalism, which has metaphysical connotations. But scientists are typically following the empirical perspective about physical world. As such they want to treat the cosmological entities -- standard model, quantum gravity and allied mystical entities – discovered by science as the reality or first cause of universe. These fundamental entities altogether is ironically described as *physical reality*, in spite of the fact that it is the subject of phenomenal knowledge only. The distinction between reality and phenomenon may be recalled here. Further scientists hold that if the cosmological entities are assumed to exist then it will serve as the justification for scientific propositions. This doctrine is conveniently called as **scientific realism**. It amounts to saying that the physical world exists just because its basic components exist. [# 6].

As per our exposition, the basic entities of quantum cosmology have been discovered by means of the methodology called mophism, which relies on the combined strength of mathematical formulae and experimental observations. In this situation, the existence of cosmological entities implies that matter and energy exist as the fundamental constituents of physical world. This kind of argument of scientific realism is normally qualified as *Inference to the Best Explanation (IBE)*. The technical expression of IBE is given below.

a) The statement P is scientifically true.
b) The fact Q gives the best explanation for P.
c) Therefore Q is true.

According to this argument it is necessary to accept that basic entities in the invisible world exist in order to explain the phenomena in visible world in the best way. Such inferences in content view are accepted for practical purposes. Moreover it is the ultimate testimony about the success of science. But the inference got in this way need not always be correct. For example: During my walk in the morning I saw water on the road. It is an IBE to infer that there was rain last night. But it may turn out to be a wrong inference. The wet road might be due to the water leaked from a tanker which passed through that road last night. This is ascertained when proper evidences are obtained.

It is the nature of our scientific mind to think that basic physical objects are real for giving meaning to scientific statements. The love, trust and conviction for knowledge depend on the assumption about the existence of the concerned objects. We like to distinguish between imagination and fact. A religious believer might feel that there is no meaning in prayer if there is no God. Although the existence of God is a phenomenal knowledge, it is a practical necessity to believe that God exists. Similarly, believers of science are inclined to accept the reality of basic physical entities. But it has already been clarified that it is a phenomenal and practical thought. The stand that basic factors of physical world exist really needs to be analysed philosophically. Now we shall see how the problem of *skepticism*, which is inherent gravely in naturalism and naive realism, affects scientific realism too. [# 7].

We may recollect the traditional form of skepticism, as appearing in the empiricist philosophy of David Hume, with regard to the question whether matter exists really. Hume was concerned with the visible level called macro world, which is made of atoms having three dimensions of length, breadth and depth. As per our present knowledge, we can say that atom is a combination of matter and energy. Accordingly, the various physical properties such as weight, color and taste subsist on matter, which is treated as a pure substance having the only property of extension. *Now it is proper to recast Hume's question as: can we know that matter and energy exist so as to constitute the physical world?* [# 8].

It may be recalled that Hume falls into skepticism about the question about matter because he has only empirical knowledge. Coming to the elementary levels of physical world -- such as Standard Model, superstrings, membranes, dark matter and dark energy – we have explained above that the fundamental entities are proposed to exist *conventionally in the form of an IBE*. Extending the arguments of Hume we can hold that the fundamental entities are mere abbreviations of respective sets of empirical properties. They are proposed to exist in order to satisfy the practical objectives of science. We can add that the scientific realism has the following defects also:

1. The *problem of induction* which disturbed David Hume is also applicable to the basic entities. The inferences about basic entities accepted now may prove to be wrong in the future through better experiments.
2. It should be kept in mind that scientific knowledge is formulated through TyHDTI scheme, which is not necessary truth. Amendment to theories is a basic characteristic of science. Due to this possibility, there cannot be an ultimate theory about basic entities of physical world. It shows that Inference to the Best Explanation (IBE) also leads to *skepticism*.

If the fundamental entities of matter and energy -- standard model particles, superstrings, membranes, dark matter and dark energy – do not exist actually, how can we say that physical world exists? What is the correct justification of physical science? The next section will apply the ground-breaking principles of System Philosophy for the visible and invisible levels of physical world.

5.4 System Model of Justification - Existence of Physical World

In the foregoing discussion, we have made it clear that justification must be achieved under *content view* because process view is interpretative without the notion of existence. This is reason for attempting to clear the issues of naïve and scientific realism. *It may be observed that both David Hume and Immanuel Kant have more or less the same position that physical world is a phenomenon that is a construction of our scientific mind. However, as a clarification, we may add that the things such as sun, earth, table, water, atom, proton and quark should not be construed as imaginary objects like angels or ghosts. On the other hand, there is an unknown reality for physical world, which is transformed into matter and energy for conceiving the physical things as phenomena for practical purposes.* [# 9] [*]

The quantum cosmology wants to explain the origin and development of physical world. If you say that there are many entities as the fundamental constituents of the world, it is a statement of *pluralism*. I can challenge it by the question: how do the pluralist entities come into existence in the first place? Can we think of reality without the ideas of symmetry and unification? Then we have to admit that pluralism cannot be the path for conceiving the integrative feature of phenomenal world.

For solving the justification problem of physical science, we must establish the following aspects:

- If matter and energy are considered separately, these are mere predicates, which cannot be said to exist – this idea follows from the philosophy of Immanuel Kant. So we have to prove that matter and energy exist phenomenally in an interrelated manner.
- The existence of matter and energy at the elementary levels must explain the formation of the hierarchy of physical world.

In this line of thought, the following treatise is intended to dispel the skepticism and pluralism related to the innumerable constituents of physical world.

Physical World is a System [*]

For achieving a breakthrough in the issue of justification, we must make use of the two principles, namely, principle of symmetry and principle of system.

In the chapter 3, we have introduced the table about the framework of quantum cosmology by linking it to the ideas of symmetry and symmetry-breaking. It may be further added here that this principle is best illustrated in the growth of a tree. Through abstract reasoning we can hold that the original seed has perfect symmetry. The first event of symmetry-breaking happens at the time of the germination of the seed to become a plant. And this process is repeated when the plant grows in the successive stages to reach the size of tree with many branches and leaves. Similar stages of phase transition are evident in the case of biological evolution; the first cells underwent symmetry-breaking and it is the beginning of evolutionary process leading to the numerous varieties of biological species. This principle can be logically applied for deliberating upon the origin and evolution of physical world.

We are primarily concerned with the physical world existing in four-dimensional space and time; the Big Bang represents its origin. In the initial period of 10^{-35} second there was perfect symmetry between matter and energy – this is the stage of *quantum gravity*. It is alternatively seen as the plasma state in which gravity and standard forces are unified. Since gravity has negative energy and standard force has positive energy, it can be concluded that the net energy was zero at this stage. Stephen Hawking (*Brief History of Time*, page 136) subscribes to this view and says that our universe originated from the state of zero net energy. [#10].

The symmetry-breaking of quantum gravity resulted in the GUT stage where gravity and standard force got separated, making the latter as predominant. Successive events of phase transitions caused the stages of inflation, Higg's mechanism, formation of elementary particles of standard model, emergence of subatomic particles together with basic forces leading to the evolution of the visible level of physical world.

The above historical perspective implies that deliberation about Big Bang and quantum gravity is the crucial part of cosmology. Theoretical physicists including Albert Einstein and his successors have struggled to develop a theory about quantum gravity that can synthesize the opposite aspects of gravity and standard force – such a theory is grandiosely named as the *Theory Of Everything (TOE)*. But it remains only as a dream since the required principle for synthesizing matter and energy is elusive to the concerned researchers. The theoretical physicists do not know the philosophy for integrating the opposite entities called matter and energy. Against this background, we can propose the method of System Philosophy as the solution for understanding the basic stages of cosmology. [#11] [*].

The most intractable question about big bang theory is: what did exist before as the cause of Big Bang? It is necessary to discard the standard version that the entire mass of present universe, which is estimated at 10^{50} tons, was concentrated at the singularity point and it got scattered subsequent to the big explosion. In the scientific parlance, **the term 'present universe' stands for our physical world in four dimensions of space-time.** As a sequel to the discussion on Superstring theory and M-theory given in third chapter, we can freshly interpret Big Bang by introducing the idea of *past universe. Accordingly, we propose that the past universe existed with 10 or more dimensions. The Big Bang is the event of compactification happened to a part of past universe by which the four dimensional present universe emerged.* Compactification is the process of hiding the extra dimensions in order to reduce into four-dimensional space-time.

In this line of argument, Big Bang is the singularity point representing the beginning of the process of compactification, which continued throughout the history of present universe. It is instructive to say that the singularity point can be compared to a seed lying in the field of past universe. Then Big Bang is similar to the event of the germination of the seed to become a plant. Note that the growth of plant to become tree happens by drawing resources like water and manure from the land. Similarly, the matter and energy belonging to present universe is obtained by the transformation of the factors of past universe. This view can explain the expansion of our universe during the past 13.70 billion years.

As suggested above, the present universe (physical world in four-dimensional space-time) has originated from a past universe having ten or more dimensions. Leaning on the example of land and tree, the existence of physical world depends vitally on that of past universe. So our philosophical investigation moves to speaking about the existence of past universe. The philosophical idea of **system** can be adopted fruitfully in this context. The term universe (A) refers to the total of past universe (B) and present universe (C). That is, we have the equation, $A = B + C$. In this situation we must show that *the universe (A) is a system with X and Y coordinates.*

In order to apply the notion of system to the total universe, we have to firstly deliberate upon the existence of past universe. Here, the recent discovery of dark matter (DM) and dark energy (DE) is extremely useful. As explained in previous chapter, DM and DE are opposite entities; DM is responsible for the formation of ordinary matter of our universe, while DE represents the antigravity force causing its expansion. Before the event of big bang, that is in the negative part of time, we have the equation $A = B$ because $C = 0$. In other words, before big bang, past universe is 100% of universe.

Coming to the existence of past universe (B) we can now postulate that it is a system composed of dark matter and dark energy -- these opposite components together make a whole. As per system philosophy, the past universe is depicted by X-Y coordinate system; where X represents DM and Y represents DE. This system model can be illustrated by a diagram, which is similar to the X-Y model mentioned earlier. The four quadrants of the diagram consist of the mysterious cosmological entities such as membranes and multiverse proposed by M-theory. In other words, the DM-DE system 'exists' as the unifying principle for the entities which are imagined to inhabit the past universe. Here the word 'exists' is used in a commonsense way. We will philosophically examine the notion of existence in later paragraphs. [*]

Scientists treat DM and DE as physical concepts, being the inferences from the scientific investigations pertaining to the origin of our physical world. But the separate existence of DM and DE can be challenged by the philosophical issues of *modern phenomenalism* and *scientific realism*, as explained earlier. In contrast, it is enlightening to note that the notion of DM-DE system confirms the phenomenal existence of past universe, without involving the drift to realism.

In the above paragraph, we have defined past universe as DM-DE system with four quadrants. However, when we consider the origin and development of present universe, it is expedient to say that there is a **pocket universe** in the third quadrant from which the present universe emerges. Hereafter, the term past universe refers strictly to this particular pocket universe. [*]

After the event of Big Bang, DM and DE are being converted into matter and energy respectively as required for the expansion of present universe, i.e., our physical world. With the advancement of time the physical world expanded so that today its mass, estimated at 10^{50} tons, accounts for just four percent of the total matter-energy content of universe. Specifically B = 96% and C = 4%.

Here it is possible to divide the straight line of time into two parts, negative and positive. The time in the present universe is conventionally taken as positive, while the time of past pocket universe is considered as negative. *The arrow of time is always in the direction from left to right — that is, time moves from minus infinity to plus infinity.* [*]. This peculiarity of time line will be explained in section 6.1 of next chapter.

The present approach conforms to our common sense and it will avoid the nagging confusion about time expressed in various books of popular science. Then, past universe (B) and present universe (C) can be arranged successively according to the straight line of time so that the event of big bang is represented by the point zero. That is, the negative part of time belongs to B while the positive time belongs to C. Now we assert that the *universe is a system of X-Y coordinates* where -X represents DM, -Y represents DE, +X represents matter and +Y represents energy. Hence, the past universe appears in third quadrant while the present universe (physical world) belongs to the first quadrant. The X-Y coordinate system is called the **System Model of physical world** according to the content view; then the point (0, 0) is the singularity point of big bang. [*].

It may be emphasized that *our present universe (physical world) is a system of matter and energy* because it is the positive part of universe as an X-Y system. The foregoing technical method of depicting universe and dividing it into two parts – past and present – ingeniously solves the problem of existence. Further, the physical world exists with two levels (the *macro* world of atoms and the *micro* world of subatomic phenomena), each of which are combinations of matter and energy. Then we can interpret past universe as the *unobservable* world constituted by dark matter and dark energy taking ten or more dimensions. In contrast, the present universe is the *observable / observed* world constituted by matter and energy in four dimensions of space-time. Note that we can observe only things existing in three dimensional space together with one dimensional time.

In this context, it is possible to provide the coherent **definition of big bang**. Of course, contrary to the popular myth, big bang does not mean any explosion. We have to invoke here the notion of compactification pertaining to membrane theory in order to get the definition: *Big Bang is the beginning of the process of compactification by which the extra dimensions of past universe are hidden resulting in the four dimensions of space-time.* That is, Big Bang is the event when our present universe took birth in the field of past universe.

Now we may reconsider the analogy of land and tree for conceiving the existence of past universe causing the development of our physical world. Alternatively, the tree may be replaced by a small circle; then concentric cycles (expanding balloon) represent the expansion as per Hubble's theory. The corresponding diagrams illustrate these two ways of conceptualizing past and present universes. The picture of concentric cycles must be modified by putting it inside a large circle denoting the past universe. So there is enough space for the expansion of our present universe. [#12] [*]

The problem of justification with regard to scientific knowledge is solved as following. Theoretical physicists propose that many levels of cosmological entities exist -- such as multiverse, strings, membranes and quantum gravity – though these are mere mathematical models. Though a person naively thinks that the justification of scientific knowledge rests on the actual existence of concerned cosmological entities, this point must be examined philosophically. We can reject this scientific realism by subscribing to the assertion of Immanuel Kant: *existence is not a real predicate*. We cannot include 'existence' among the set of theoretical properties attributed to various levels of cosmological entities representing matter and energy. Accordingly we postulate that it is a *theory* when we say the statement: *physical world is the system of dark matter (DM) and dark energy (DE)*. This DM-DE system rightly becomes the so called *physical reality* meant for describing the hierarchy of physical things. Since DM-DE system is a theoretical idea, it does not imply existence.

The meaning of a scientific proposition must be based on the view that it is a combination of DP and IP involving the stages denoted by Ty, H, D, T and I. That is, the propositions expressing scientific knowledge are *inferences* formed by this scheme. Further, for avoiding realism, we accept that the fundamental cosmological entities are mere predicates. That is, the entities of cosmological levels are phenomenal descriptions of DM-DE system; these are theoretical ideas without implying existence. It is envisaged that the dialectical and productive relation between DM and DE produces various forms of matter and energy. The existence of such products is an inference based on evidences. This leads to an important point: ***Existence is an inference; it should not be treated as part of theory***.

The above arguments assert that the existence of physical world (DM-DE system) is also an inference. Those who are unhappy about the removal of realism can be convinced as following: The system model – the symmetric structure of opposites – is rationally conceived for understanding the phenomenal existence of physical things. The proof of the pudding is in the eating. ***And, we can practically say that physical world exists as per system model.*** All things of physical world exist as systems of the opposites namely matter and energy. But we ordinarily and conventionally say that individual entities like matter, energy, DM, DE, electron and quark have existence, by the application of Aristotle's rules of thought. [# 13][*].

As per the foregoing, we can conclude that the *System Model of physical world* effectively solves the issues of justification with regard to our scientific knowledge. It is now clear that System Philosophy can be applied to articulate the epistemology of all levels of physical science namely classical science, quantum mechanics and quantum cosmology. The resulting treatise is named as *System Philosophy of Science,* which can be elaborated later to include the theory of knowledge under biological sciences and social sciences also.

5.5 Spiritual Science of Ancient and Medieval Periods

Having explained that the origin of *science* is from the Renaissance happened during fifteenth century, we may briefly go back in history to explain the characteristics of certain subjects that inherently combine the aspects of science, religion and art. We can identify such peculiar disciplines as **spiritual science** noting that this paradoxical phrase refers to the traditional forms of knowledge about natural world, which were highly influenced by the religious beliefs of ancient and medieval periods. It is a mixture of rational ideas and mystical thoughts; so we can see therein the mélange of content view and process view. The wide area of spiritual scienceis divided into *mystical process science* and *pseudo science,* for reasons to be given in the ensuing paragraphs. [# 14] [*].

It is a curious fact that mystical process science and pseudo science are still accepted by some groups of people in the present period in spite of the spread of modern science. There are different versions of these esoteric disciplines in the context of western and eastern societies;

but we will concentrate only on the epistemological aspects in concise manner.

Mystical Process Science

Let us first consider the traditional forms of western knowledge that combined the thoughts of organic worldview (idealism) and spiritual process worldview (pantheism) as well as the beliefs of mystery (pagan) religions pertaining to ancient Greece and areas of Roman Empire. The rational principle of such knowledge is the legacy of Aristotle who was the main philosopher to introduce the scheme for the systematic study of world through content view.

According to Aristotle, the changes happening in concrete things and their relations would manifest certain good purpose or teleology. In accordance with his *theory of four causes,* the worldly phenomena are analyzed using the notions of material cause, efficient cause, formal cause and final cause. Consider the example of the production of a cup out of the lump of clay, by a potter. The lump of clay serves as the material cause. The technology including the instruments, energy and labour of potter forms the efficient cause. Thirdly, potter has the design of the cup in his mind while engaging in the production work. This design is the formal cause. Fourthly, there is a final cause meaning the purpose or aim for which the cup is made. Aristotle held that the efficient, formal and final cause can be clubbed into a class called form. Thus the substance of cup is a combination of matter and form.

Form is in the nature of thought or idea and it determines the actuality of a substance. Matter or the physical stuff is a mere potentiality, to be molded according to the form. The hierarchical order of things in the universe can be regarded as higher levels of form – it is Aristotle's notion of teleology. In this way, Aristotle argued that God must exist as the totality of forms and the prime mover of formless matter. Reality of world is God as supreme form or pure mind.

For studying the cause-effect relations of natural world, Aristotle deals with material and efficient causes expressed as qualitative aspects of phenomena. Obviously it is based on the notions of mental and material substances. Material and efficient causes are defined in terms of rational abstract properties of species, without observing the empirical (sensible) aspects of particular objects. God and allied supernatural forces accounted for the design (formal cause) and purpose (final cause) of things. The notion of vital force (*élan vital*) was employed to explain the life and mental activities. So a particular medicine represents only the material and efficient causes, while its curative power really depends on the metaphysical factors such as vital force and divine will.

Generally speaking, the rational part of *mystical process science* consisted of the deductive propositions pertaining to the material and efficient causes; these were assumed to be based on the notion of formal and final causes coming from God. Aristotle proposed that the knowledge about material world consisted of deductive propositions which exist in the form of syllogism: *premise-fact-conclusion*. Here, the premise is derived from a metaphysical theory about the relation between supernatural forces (God is assumed to be its supreme form) and material world. Fact is the cause-effect relation like 'cloud causes rain' observed in the natural world. Then, for example, a typical deductive inference is given in three stages as under:

Premise – God created matter and He controls everything in this material world.

Fact – Cause-effect relation like 'cloud causes rain'.

Conclusion (deduction) – Cause-effect relations of world manifest the will of God.

The most famous achievement of Aristotelian science, which involved both idealism and mysticism, is the astronomy of Ptolemy who lived about 150 AD. Applying the method of geometry, Ptolemy postulated that planets, sun and other stars moved around a fixed earth as centre. This mathematical astronomy was wedded to the astrology, which is the religious interpretation of the heavens – it held sway up to the fifteenth century. The divine qualities of stars and planets combined with the astronomical computations provided the science-religion interface for explaining the phenomena on earth. This geocentric theory ruled western thought till 16th century, when Copernicus (1473-1543) replaced it with his heliocentric theory. Through ingenuous experimental observations, Copernicus showed the fallacy of the premises and deductions of Ptolemy.

Additionally, astrology had heavy influence on the western societies of ancient and medieval periods. The calculations about the movements of planets and stars were associated with the mythology of pagan gods so as to interpret the events in human life. Further we have to mention about the ways of medical treatment. The system of medicine developed by Hippocrates, the Greek physician of fifth century B. C., was in the line of *mystical process science*, because the parts of body were examined as combinations of material and mental aspects. Subsequently, in second century, the researches of Galen contributed many ideas about the functions of different parts and organs of the body. His theories were accepted for many centuries and served as the precursor to the modern medicine called Allopathy. It is worth noting that Hippocrates is regarded as the father of modern or western medicine.

The process worldview embedded in the doctrine of *pantheism* had a special role in the mystical study of biological organisms and medicine during ancient and medieval periods. It promoted the development of homeopathy in Germany during 18th century and this alternative medicine is now practiced world over.

In the eastern world, the main versions of spiritual science are *Ayurveda, yoga, naturopathy, pranic healing and acupuncture.* These are traditional medical systems following the spiritual process worldview or pantheism; they place more emphasis on deductive knowledge as compared to induction. In order to highlight the concerned issues, we may take up the raging **controversy whether *Ayurveda* can be treated as science or not.**

Ayurveda considers human being as a system of body and soul (consciousness), following the Indian philosophical doctrines mainly

Sankhya and *Vaiseshika*, which have strong undercurrents of pantheism or spiritual world view. It means that Ayurveda does not analyse human body in a physical manner as done in biological science. As per Ayurveda, human body is made of five gross elements (*panchabhutha*) as defined below:

- Space or *akasa* (associated with expansiveness)
- Air (associated with gaseousness, mobility and lack of form)
- Fire (associated with transformation, heat and fire)
- Water (associated with liquidity and instability)
- Earth (associated with solidity and stability)

Ayurveda names three principal qualities or three *dosha*s (called Vata, Pitta and Kapha), and states that a balance of the doshas results in health, while imbalance results in disease. Vata is composed of the space and air elements; it controls the nervous system. Pitta is composed of the fire element and it rules the digestive, chemical and metabolic function. Kapha has its main seat in the stomach; this dosha is related to mucous, lubrication, and carrying nutrients into the arterial system. Kapha also governs immunity; its energy promotes the ongoing processes of self-repair and healing. Ayurvedic physicians believe that the imbalance of a particular dosha produces corresponding symptoms, which are different from the symptoms of another dosha imbalance. For example, if the person aggravates pitta, he/she may develop prickly rash or an acidic stomach. Many factors can cause imbalance, including a poor diet, too much or too little physical or mental exertion, chemicals and germs.

My main arguments to hold that ***Ayurveda is not science*** as per the strict definition of this chapter are given below: [# 15][*].

- It can be commented that these five gross elements and three doshas are metaphoric concepts and they do not fall into the structure of physical world adopted in classical science. It is impossible to bring them into the framework of the scientific theory of atoms. In other words, the five gross elements and three doshas are not *physical* entities as per our definition.

- It is clear that the main theoretical entities of Ayurveda are defined at the gross level of individual person. These are qualitative concepts, not quantitative or measurable entities. More over Ayurveda holds that every part of human body is a union of material aspect and mental aspect. These facts clearly serve to distinguish Ayurveda from biology and allied medical science.
- The experimental method described as TyHDTI scheme is not properly practiced in Ayurveda.
- We may note here that the meaning of science (*sastra*) has changed when we compare ancient-medieval period with modern period. In this chapter we adopt the modern meaning of science, including physical science and biological science, based on materialism. We can conclude now that Ayurveda is an example of *spiritual science* under the spiritual process worldview or pantheism.

The position of other eastern systems of medicine is lower than Ayurveda because they give still less importance to induction. The efficiency and reliability of treatment cannot be estimated satisfactorily due to the mixing up of physical (material) and nonphysical aspects in the TyHDTI scheme.

As indicated above, there is an appearance of agreement between spiritual science and religion. We can say that the harmony between science and religion under organic and spiritual worldviews – with ontological doctrines of idealism and pantheism respectively - was an illusion, characteristic of pre-scientific era. Though the syllogism employed in ancient forms of study is said to have necessary truth, it does not agree with our present definition of *physical science*. More specifically, the truth of premises cannot be ascertained through sensory observations or experiments.

Pseudo Science (Paranormal Phenomena)

Many people in various parts of world believe in the existence of paranormal phenomena, which violate the physical laws of nature. Such phenomena are alternatively called supernatural, occult or psychic phenomena; the term "psi" is used to refer to them generally. The main types of psi are *telepathy, clairvoyance, precognition and psycho kinesis as well as the practices like astrology, occult, miracles, palmistry, numerology, vastusastra and faith healing.* Telepathy means one's awareness of the thoughts of another person. Clairvoyance is the information about objects and events hidden from the senses. Precognition is the knowledge of the future unmediated by sensory inputs and rational thought. These three kinds of knowledge together is usually designated as extrasensory perception (ESP). Psycho kinesis means the direct influence of mind over matter; for example bending of an iron bar using mental power. All kinds of psi lie outside of the premise of physical causation. The ancient Indian tradition claims that a person can acquire paranormal abilities through appropriate practice of meditation *(yoga)*.

It is normally accepted that paranormal phenomena or psi are closely associated with the rituals and practices of religion; this idea is based on the fact that all kinds of supernatural events are experienced by our mystical mind, which is to be distinguished from our scientific mind. Many mystical groups or cults seek the benefits from paranormal activities. However, we can clearly demarcate psi from religious events and mystical process sciences as following:

- ➤ In religion, a believer aims to attain mystical experience of deity or object of worship. The religious practices represent good purpose of life and after-life. On the contrary, the activities of psi have practical purposes, mostly of the bad kind. Generally speaking, the experiences of psi are motivated by the evil intentions of mind such as anxiety, greed, jealousy, enmity, exploitation and irrational doubt. The practitioners of psi are generally greedy so as to exploit credulous people.
- ➤ We have seen above that the mystical process science combines the aspects of science, religion and art. But such disciplines contain the good purposes, mainly curing diseases and attaining mental health. This point serves as the best criterion to see that psi is a different category altogether from mystical process science.

In last century, many researchers in western countries tried to prove the existence of paranormal phenomena by conducting special experiments – this area of research is known as ***parapsychology***. But there is a general criticism that the experiments of parapsychology are contrived and manipulated, since the supernatural phenomena cannot be captured by physical equipments. It must be understood that all types of psi are peculiar subjective experiences of our mystic mind; these are outside the realm of genuine scientific method. We cannot conduct repeated experiments in laboratory under fool-proof conditions for testing the features of psi. In this situation the phrase **pseudo science** became popular in order to despise the experiments of parapsychology. Though there are many professional research groups, institutions and journals in the field of parapsychology, they have failed to demonstrate the laws of psi in a convincing manner. [# 16].

Additional Points about the Status of Spiritual Science

The philosophical deliberation about spiritual science is a relatively uncharted area; I may give some original ideas on this topic. There are two important aspects that we have to consider now:

- The limitations of physical science in dealing with the various spiritual phenomena, more specifically the paranormal events.
- The epistemology of spiritual science, with the branches called *mystical process science* and *pseudo science*.

As mentioned earlier repeatedly, science is the systematic study of physical phenomena conceived in the framework of matter-energy or space-time. But theoretical physics has not succeeded in revealing the secrets of nature; cosmology resorts to mathematical models, which cannot be treated as proper information about the reality of physical world. Most importantly, science fails in explaining the origin of consciousness or mind in living beings.

On the other hand, the method of spiritual science does not satisfy the actual features of natural science because of the assumptions about supernatural forces. In spite of that, there is a group of scientists and writers who profess that the area of spiritual science is to be included in mainstream science; they want to broaden the meaning of science accordingly. We can summarily reject this view noting that such proponents have not properly understood the philosophical distinction between science and non-science.

Coming to the methodology of spiritual science, the concerned inferences are formed by TyHDTI scheme, where theory involves the mixing up of physical (material) and nonphysical aspects of nature. In this situation, experiments cannot be conducted in purely objective and quantitative terms. Further, there is more radical difference between natural science and spiritual science, with regard to the aspect of source (philosophy of mind). An original idea to be introduced in the next book is that our mind has two principal departments, namely scientific mind and mystic mind. As per this division of mind, the source of spiritual science is *mystic mind* mainly and it has rational and empirical faculties. The cognitive function of mystic mind also involves the combination of rational and empirical ideas, though it essentially has the shade of metaphor and symbolism. The role of scientific mind is very minor in this kind of knowledge.

What is the justification of knowledge pertaining to spiritual science? Following the arguments given earlier, we do not subscribe to metaphysical realism in the case of supernatural forces. The alternative way of justification is to apply the *System Philosophy of Seven Life Systems* to be developed in the third book. We formulate the propositions of spiritual science, when we participate in certain activities – traditional medicines, prayers, astrology, palmistry, etc - that belong to the interface of certain social systems, which are mainly related to nature, economy, religion and art. Such social systems are technically called as *life systems* and they have global existence as per system model. The existence of life systems would serve as the justification of knowledge under mystical process science and pseudo science.

More interestingly, the justification of parapsychology can be established in this innovative manner. For illustrating the above point, let us consider the knowledge related to astrology. For example a credulous person seeks the predictions of an astrologer. This interaction is in accordance with the religious belief of the person, suppressing the scientific spirit. There is a significant element of art and money in the performance of the astrologer. So the activities of astrologer and client happen in the life systems of religion, art and economy. The existence of such life systems as per system model is the justification of the inferences of astrologer.

It is a well known fact that there are evidences about the occurrence of various paranormal events, collectively denoted by psi. Even animals and birds display the psychic abilities of premonitions and precognitions. They normally anticipate earth quakes and other natural disasters. In the case of human beings, paranormal phenomena are the products of our *mystic mind* that can mimic the activities of scientific mind. This fact specifically explains the phenomenon of possession. As an example, certain person is said to be possessed by evil spirits. In some cases, it appears that particular spirit of the dead speaks through an affected person.

Generally, there is no validity in saying that paranormal phenomena are illusory. Since we can establish the justification of psi using the principle of *Life Systems*, the derogatory phrase *pseudo science* is unwarranted. Parapsychology is a form of nonscientific knowledge that can be studied through the method of System Philosophy. Anticipating the theory of truth to be developed later, we can state that the concerned propositions about paranormal phenomena have a combination of necessary truth and contingent truth.[# 17].

NOTES of Chapter 5.

1   Main references about philosophy of science are: Bird, Alexander (2003); Grayling (Editor) (1995);Martin Curd and J. A. Cover

(1998); Newton, Roger (2010); Rosenberg, Alex (2000); Urmson J.O. and Jonathan Ree (1989).

#2 The main introduction of System Philosophy is postponed to next book, where we discuss the phenomenon of life.

#3 The phrase "holism of meaning" is based on the idea of W. V. O. Quine (1908-2000), as mentioned in Rosenberg (2000), page 151.

#4 Paul Feyerabend articulated this view in his book *Against Method* published in1975. It may be noted here that Paul Feyerabend and Thomas Kuhn are dealing with change of paradigms due to social conditions. But Karl Popper was concerned with the methodology of science under content view within a given particular paradigm. Recall his falsification theory mentioned in section 4.3.1.

#5 An important development following Kuhn's approach is advanced by Imre Lakatos in his book *The Methodology of Scientific Research Programmes* (1977).

#6 Scientific realism is a consequence of materialism. Idealism, which is the doctrine of reality opposing materialism, holds that the ultimate cause of universe is Mind and it is the source of ideas. Plato and Aristotle are the original exponents of idealism; according to them God is the supreme form of mind.

#7 Here we consider the notion of God under theism which is a branch of philosophy of religion. The theist God is a person having infinite power, goodness and perfection; it is the object of prayers and worship. The philosophy of pantheism holds the alternative view of God as the basic force and creativity immanent in nature – this mystic view is not directly linked to religious worship.

8   The issues about the empiricism of David Hume are discussed in Rosenberg (2000), Anthony Harrison – Barbet (1990) and Grayling (Editor) (1995).

9   Kant, Immanuel (2003), *Critique of Pure Reason*, translated by J. M.D. Meiklejohn (Dover Publications, New York, 2003), Transcendental Dialectic, Book II, Chap. III, Section IV, pages 331-336. The critical philosophy of Immanuel Kant is adapted here for explaining that the fundamental entities of quantum cosmology are mere predicates. (This is my original idea)

10   Hawking (1995), *A Brief History of Time*, Bantam Books, 1995 edition, page 136.

11   Michio Kaku and Jennifer Thompson (2007), pages 4-11. We have already mentioned in previous chapter that the superstring theory has many drawbacks; hence, the synthesis of matter and energy has not been achieved. (this is my original idea)

12   The balloon model of expanding universe is explained in Green, Brian (2005), pages 231-238. We can say that for the expansion of our present universe, there must be space outside belonging to past universe. (This is my original idea)

13   These important points together with the system model would remove the problem of agnosticism prevailing in Kant's philosophy.

14   Using the paradoxical phrase *spiritual science*, which is my original idea, we can clear the ambiguities about the nature of ancient knowledge about worldly phenomena.

15   The historical change in the meaning of the word 'science (*sastra*)' is the main reason behind the raging controversy whether *Ayurveda* and *Yoga* can be treated as science or not. In

this chapter, we treat science as the knowledge under physical view about material aspects of things.

#16 It is paradoxical to say that the functions of nonphysical forces can be studied through physical experiments. However, in parapsychology, a lot of experiments have been conducted adopting certain procedures which do not strictly conform to scientific method. See the following articles:

- *Paranormal Phenomena and Science* by Augustine Perumalil published in Job Kozhamthadam (editor) (2004), pages 149-174.
- *In Search of Unity: The Meeting of Science and Spirituality in Mind Sciences by* K. Ramakrishna Rao in Job Kozhamthadam (editor) (2005), pages 145-174.

17. Sheldrake, Rupert (2013), *The Science Delusion*, pages 231-259. Sheldrake's treatment of paranormal phenomena is rather descriptive. Since he does not delve into the epistemological aspects, he fails to answer the question whether paranormal phenomena are illusory.

Chapter 6

The Cosmological Puzzles Finally Solved

****

6.1 Space, Time and Gravitational Waves

6.2 Solution to the Dilemma about Matter.

6.3 FAQ about the Origin of Universe

6.4 Do We Need Intelligent Design Argument?

Author's main original ideas are marked by [*].

The mark [#] gives the number of note at the end.

****

In the preceding five chapters we discussed various problems in the philosophy of science, especially those related to cosmology. There is no doubt that various aspects of cosmological puzzle, which we encountered in chapters 2 and 3, are the most important bones of contention in the entire field of scientific enquiry. And, the hitherto development of philosophy of science has been inadequate and weak in solving the mysteries connected with the experimental research about the origin and evolution of physical world. In the present chapter, we would consider the list of main cosmological puzzles which are in the upper mind of every scientist, theologian and ordinary person. It can be recognized that the tenets of System Philosophy with regard to the method and justification of physical science are uniquely suited to propose credible answers.

Nobody can venture into the study of cosmology without the utmost sense of wonder, because the size of universe is infinitely large. It is scientifically known that during the period from 1 billion years to 9.10 billion years after Big Bang, about 100 billion (10^{11}) galaxies came into existence. As a further estimate, one galaxy contains 10^{11} stars. Milky Way is one among the galaxies and it includes our sun and the solar system containing our earth. It is awe-inspiring that the present universe has the diameter of 10^{23} kilometers and it contains 10^{50} tons of matter.

The first puzzle challenging us is to explain the source of such unimaginable quantity of matter. It is now possible to tide over the controversies about the singularity point of origin pertaining to Big Bang theory. Our description of the theories about membranes, compactification, dark matter, dark energy and past universe are the key ideas in the great saga of universe under physical view. For accounting the existence of matter and energy, we have to make use of the notion of complimentarity as well as the system model of physical world.

In order to complete our treatise about matter, it is necessary to examine the prevailing confusions and alternative interpretations. Hence, in the ensuing sections, we may deal with the controversies about space, time, gravitational waves and other fundamental aspects of physical world; finally we would take up the *intelligent design argument* also.

6.1 Space, Time and Gravitational Waves

Three fundamental concepts about physical world, namely *space, time and gravitational waves* will be discussed here from philosophical perspective. Space and time are the most important theoretical concepts for all branches of physical science, especially cosmology. On the other hand, the idea of gravitational waves is an interesting phenomenon in the area of astrophysics. This article aims to consider the various definitions and meanings given to the said terms in different theoretical situations. The principles of System Philosophy can be applied to this context for solving the conceptual dilemmas.

Astrophysics, which is the scientific study of the astronomical bodies like galaxies, stars and planets, give the experimental information required for developing the cosmological doctrines. We may note that the astronomical bodies were formed during the long period between 1 billion years and 10 billion years after big bang. The phenomena of space, time and gravitational waves are fundamental for the understanding of cosmology and astrophysics.

The most basic theoretical concepts pertaining to all levels of physical world are *Space* and *Time*. As such in the ensuing analysis, space and time will be firstly considered concentrating on macro and micro worlds. Subsequently, its cosmological aspects and the notion of *gravitational waves* will be taken up for philosophical analysis.

Space and Time

It is a common sense view that space is like a container of things in this nature. To clarify this point, consider a room with air and ten chairs only. If the air and all chairs are taken out of the room, we say that only space remains in the room. In which sense can we say that space exists within the four walls of the room? Is it true that **space** exists as a physical object so that it can be used to place things like table, chairs, air and water? Generalizing these questions to the whole of physical world, the concept of space raises more serious issues. Is it justified to think that space exists initially in order to contain galaxies, stars, sun, moon, earth and so on through the evolutionary history of world? *We are perplexed by the idea that space exists as a physical substance in the form of a container and at the same time it represents the absence of all physical objects.*

Galileo and Isaac Newton, who developed the classical physics, elaborated the above common sense view. They treated space and time as independent, absolute and real entities in accordance with the mechanistic worldview. We can explain this idea in a simple way by considering a box having three balls and remaining empty space. Then the space inside the box has a particular structure at time T1. Let us change the position of balls in the box. As a result, the structure of space inside the box has undergone change at time T2. But it is true that when space inside the box changes it does not affect time; time moves from T1 to T2 in a uniform manner. In other words we can change the structure of space without changing the flow of time. Observing the physical world as a whole, we can note that the astronomical objects are in continuous motion whereby the structure of space alters along with uniform flow in time. Though change of space always involves the change of time also, the classical science tried to treat that space and time as independent and absolute entities in the macro visible world.

It is possible to define 'time' simply as a concept used for understanding the occurrence of various events in physical world. **Time** is the interval between two events related by the notion of before and after. We can note that the idea of cause-effect contains the concept of time also; the cause always occurs before the effect. For example, cloud causes rain. The cause (cloud) occurs before rain (effect). When the events of world are linked as a sequence of cause-effect relations, it becomes a sequence of before-after. If we represent the concept of time by a straight line, it includes the occurrence of cause-effect relations. Then, the timeline runs from left to right; this aspect is called the ***arrow of time***. Obviously, the events in physical world do not occur backwards in time because the reversal of time is not allowed by the principle of cause-effect. [# 1] [*]

Let two points A and B on timeline represent two events so that A occurs before B. The distance between these two points is called the time interval between the events. The important issue here is to divide the time line into equal intervals according to a scale. It is achieved in the history of human civilization by observing the regularity of events pertaining to the movements of earth, moon and sun. For example, sunrise was observed at regular intervals giving the notions of day and night. Let us define the word 'hour' so that the time duration between two consecutive sunrises is taken as 24 hours subject to certain generalized experimental conditions. Using this definition of hour, we can define the aggregates like week, month and year. Similarly, we can divide hour into its parts like minute and second. On the basis of these concepts, the straight line of time can be marked with appropriate units. The point being emphasized here is that time is a concept only, which is derived from observing two regular events where one is after the other. However, classical physicists assumed *naïve realism* to treat time as an object like river which flows in the direction from left to right.

In the early decades of last century, Einstein presented the revolutionary *special theory of relativity* according to which space and time are relative or interdependent. It was expressed using mathematical language as well as thought experiments. *Einstein's theory was intended to explain the motion of subatomic particle which possessed very high energy (speed) in contrast to their extremely small mass.* We can remark that the space-time relativity is the underlying principle meant for explaining the particle-wave duality of subatomic phenomena. So it applied specifically to the world of quantum mechanics, but was subsequently projected to explain the macro world also resulting in the overthrow of mechanistic worldview.

Einstein adopted the traditional view of naïve realism, which is later extended as scientific realism, implying that space and time are really existing inter-related objects; this view opened the flood gates for many fantastic forms of thoughts and controversies like *quantum space and time*. So far we have seen that the notions of space and time are different in classical physics and quantum mechanics. Our methodological problem is to obtain a coherent interpretation for the levels of meaning of these concepts. [# 2].

It is required now to reconsider the *System Model of physical world* developed in previous chapter. Thus we have convincingly solved the issues of ontology and justification with regard to our knowledge about physical world. The main points relevant in this context are restated below.

The notion of absolute existence of theoretical entities causes a dilemma, since it holds that space and time began at the event of big bang. If our universe originates at the point of big bang and it expands with advancing time, we are prompted to ask: how can the space of our universe expand if there is no space outside it? For answering this question, we must enquire whether we can treat space and time as really existing entities of physical world.

Referring to Immanuel Kant (1724–1804), we can refute realism by holding that the concerned theoretical entities of science are *predicates*, which are defined by certain properties. Such properties are conceptions of mind with or without the support of empirical data. In any case, the predicates included in theory are deductive concepts pertaining to our phenomenal knowledge. According to this line of thinking, the theoretical entities of physical science cannot be said to exist really. This principle can be gainfully employed in the case of *space* and *time*.[# 3].

In this way, we assert that *space is a theoretical idea required for describing the phenomenal existence of physical objects*. Further, the time scale is constructed traditionally by observing the movements of earth, moon and sun. Our familiar idea of time is of local nature and it is used for the study of the events in our life as well as the movements of distant astronomical objects like stars and galaxies. Here we have to avoid realism; time does not exist as an object. *Time is a theoretical concept representing the abbreviation of the regularity of two events*. So we admit that space and time are theoretical entities; they give the prior structure of thought for designing experiments and getting sensory experience from them. [# 4][*].

Now let us extend this idea to the area of quantum physics. The theoretical entities (predicates) like space, time, matter, energy, gravitational waves, black hole and strings cannot be said to exist really. Once we recognize that the space-time relativity is a mathematical version of the particle-wave duality pertaining to subatomic phenomena, the **justification** of such concepts requires deeper thought. In this situation, the previous chapter has advanced a new theory of justification under the heading "Physical world as a system". It holds that scientific theories and laws are justified by the phenomenal existence of physical world as a system of dark matter and dark energy as opposite entities.

Now we can remove the paradoxes and anomalies with respect to the idea of **space and time** by accepting various notions according to the levels of physical world. In the premise of system philosophy of science, we can propose four notions of space and time – *visible, abstract, quantum* and *astrophysical.* These concepts are introduced below based on the layered perspective of physical world. [# 5] [*].

Visible space and time pertains to the physical world of atoms and higher substances studied through the laws of classical physics under mechanistic worldview. Here, space and time can be seen as separate and independent entities, by the application of Aristotle's rules of thought. The idea of space must be linked to the levels of matter discussed earlier. ***For our practical purpose, the term 'space' refers to the visible world of atoms and higher substances*** [*]. The space occupied by an atom is extremely small. For example, the diameter of hydrogen atom is 10^{-13} centimeter and that of gold atom is 10^{-6} centimeter only. In the case of subatomic particles and their constituents, physicists use mathematical models of space in order to describe their motion; these models need not create any confusion about our notion of space pertaining to macro world.

Abstract space and time represents our intuitive notion of time running from minus infinity to plus infinity. The big bang theory with suitable modifications is accepted as the doctrine about the origin of our visible world. Accordingly, the origin of space and time occurred at the event of big bang happened 13.7 billion years ago. There is a serious defect for big bang theory since it does not answer the question: what did exist before big bang? The proposal that space and time did not exist before 13.7 billion years ago leads to serious dilemma. [# 6].

To tide over this issue, we extend the idea of time backwards from big bang (t = 0) and accept the notion of *negative time* with regard to our past universe. The invisible and *mysterious entities* like superstrings, membranes, multiverse, dark matter, dark energy can be used to arrive at the notion of *past universe* and it would answer the question as to what existed before big bang.[*].

It is reasonable to think that time is like a straight line of infinite length. Then, we can practically treat the big bang event as representing the middle point zero. In other words, abstract time is the sum total of the negative time pertaining to past universe and the positive time of our universe. It may be reiterated here that *the negative time runs from left to right* because we can imagine about cause-effect relations in the stage of past universe also. [*]

Quantum space and time is associated with the particle-wave duality in subatomic world; this is the notion of time included in the mathematical description of quantum mechanics and cosmology. The highlight of such phenomena is that we cannot experience space and time separately; it is the consequence of particle-wave duality. The relative existence of space-time in quantum world perplexed theoretical physicists so much that many fantastic models are suggested for its illustration. **Time dilation, time warp and time travel** are interesting descriptions of certain aspects of quantum time. Einstein's thought experiment called *twin paradox* and the explanation of time dilation are impracticable in macro world, but it is conceived for showing the space-time relativity that is assumed to exist really in subatomic world.

It is a queer idea to imagine that subatomic particles travel backwards in time. Such fantasies can be dismissed as worthless because it does not agree with the concept of time adopted for our practical purposes [*]. The principle of space-time relativity is a formal method for describing the particle-wave duality in the case of subatomic phenomena. Quantum time should not be construed as real since it belongs to the realm of mathematical models. In the context of quantum cosmology, we can point out that the authors of theoretical physics have committed two kinds of mistakes as following [*].

- They did not see the distinction between macro world and quantum world for conceiving the notions of space and time. Visible time, abstract time and quantum time are different paradigms of knowledge about time; but only the abstract time has validity in cosmological enquiries.
- They tried to relegate visible time in an effort to apply quantum time for interpreting the visible phenomena with which we are

familiar every day. It was a quixotic venture involving fantastic ideas.

Astrophysical space and time: As explained earlier, Einstein's formal model of space-time curvature is the bedrock of the theory of astrophysics. The notions of space and time would undergo drastic change in this theory and it involves the related concepts like Black Holes and Gravitational Waves also. But we want to avoid the drawbacks of scientific realism; then our system model provides the best alternative. We can depict space as X coordinate and time as Y coordinate for showing the space-time relativity. The movement of all astronomical bodies can be studied in this framework.[*].

Gravitational Waves

The phenomenon of *gravitational waves* has attracted the attention of global media and achieved wide publicity in the recent years of the present decade. It is expedient to enquire about the reasons for this development. In retrospect, Einstein had originally predicted the existence of gravitational waves on the basis of his *General Theory of Relativity* published in 1915. As per the subsequent development of astrophysics, the study of gravitational waves is now intimately linked to the theory of black holes and also to the phenomena like supernovas, quasars and dark energy. For understanding the philosophical issues of this topic, we have to consider a few points about Einstein's formalism and realism also.

We may recall the Newton's theory that gravitation is a force that takes place in the realm of space and time pertaining to mechanistic world. But, through the *Special Theory of Relativity* of 1915, Einstein had showed that space and time have relative and interdependent existence, which is called as space-time. Further in his *General Theory of Relativity*, he drastically modified Newton's view and defined gravitation as the effect of the curvature of space-time. To explain this point, we can compare space-time to a rubber sheet on which a heavy metal ball is placed. The weight of ball causes curvature on the surface of the rubber sheet. Similarly space-time becomes curved near the massive objects, mainly stars and black holes, and this phenomenon is manifested as gravitation.

Stephen Hawking writes in *The Grand Design* (page 130): "But gravity, in Einstein's theory, is not a force like other forces; rather, it is a consequence of the fact that mass distorts space-time, creating curvature". Einstein translated the traditional idea of gravitational force into the formal model of space-time curvature adopting the mathematical way. Here we must recognize that, though space-time fabric is a mathematical model only, Einstein took it in real sense. [# 7][*].

Further in a paper published in 1916, Einstein proposed that, when massive bodies move on the fabric of space-time, ripples will be generated. Such ripples on the surface of space-time curvature are called *gravitational waves*. But, how can we detect the existence of such undulations in the form of waves? Brian Green writes:

> "In principle, you could detect a gravitational wave's passing by repeatedly measuring distances between a variety of locations and finding that the ratios between these distances had momentarily changed".[# 8].

Since gravitation is an extremely feeble force in experimental conditions, astrophysicists construed that the movement of very large mass is required to produce detectable level of gravitational wave. The most feasible method is the study of movement and collision of massive black holes. With this objective, *Laser Interferometer Gravitational-wave Observatory (LIGO)* was set up in USA with the collaboration of over thousand scientists from various countries. On 14 September 2015, LIGO at first detected the signals about the collision of two giant black holes with 36 and 29 solar masses respectively. This event of collision had happened when our universe was in the very early stage that is about seven hundred million years old after big bang. (See Table 1 in section 2.2 for estimating the size of universe at that time). As a result, one combined black hole of 62 solar mass was formed. Thereby,

3 solar masses had been converted into energy, which is manifested as gravitational waves. [# 9].

What is the worth of LIGO experiment in the context of cosmology dealing with the question about the origin of universe?

Many enthusiastic scientists claim that the success of LIGO experiment has tremendous importance for the cosmological study of universe; some hail it as "the discovery of the century". We may note that similar exaggerated propaganda was unleashed in the context of LHC experiments and discovery of Higgs boson also. Maximum publicity is being given through various media in favor of the highly expensive pursuits of scientific cosmology. Here it is necessary to make an assessment of the benefits from LIGO.

Of course, we can admit that the LIGO experiment provides valuable contribution to the theory of black holes, which must be linked to the aspect of dark matter, as discussed in third chapter (section 3.3.3). It may be reiterated that dark matter forms 23 percent of total mass/energy, while ordinary matter is only 4 percent and the remaining 73 percent is dark energy. In this situation, it is reasonable to hold that black holes are major depositories of dark matter. It was further observed that Einstein's model of space-time curvature was constructed in the realm of total universe including dark matter and dark energy.

For a philosophical assessment of the benefit from LIGO experiment, it is necessary to take a critical view about the contemporary state of cosmological research. We may mention below the key points in this regard on the basis of third chapter pertaining to quantum cosmology.[*].

In the context of the scientific knowledge about physical world, the basic and crucial questions to be answered are the following:

- ➢ What is matter?
- ➢ How does matter originate and evolve to become a hierarchy of physical things?

Theoretical physics in the latest form envisages that matter has five evolutionary stages as shown below (see table 1 of chapter 3).

A. Invisible level

I (First level): *Mysterious entities* like superstrings, membranes, multiverse, dark matter, dark energy and so on. This level I includes the entities prior to big bang.

II (Second level): It consists of the earlier phenomena like Higgs mechanism, inflation, superstrings and quantum gravity and the event of big bang. It is the period up to 10^{-4} second after big bang (ABB).

III (Third level): It consists of fundamental particles of *standard model (quantum field theory)*, emerged during the period from 10^{-4} second to 1 second ABB.

B. Visible level

IV (Fourth level): **Micro world** - Four subatomic particles and four basic forces. It originated in the cosmological period from 1 second to 10 seconds after big bang. This is the area of *quantum mechanics*.

V (Fifth level) : **Macro world** - Substances made of atoms (inanimate substances including *astronomical bodies* as well as the bodies of living organisms). *Classical science pertains to this level only.* Macro world emerged in the cosmological period from 10 second to present (13.7 billion years after big bang).

We can state that levels I, II and III are successive stages of *quantum cosmology*, that is the theory about the origin of physical world. In contrast, levels IV and V pertain to the structure of physical world. The moot point is whether it is plausible to conceive the origin and evolution of matter in this way. From the details given in chapters 3 and

4, we may state below the **glaring drawbacks** of the above doctrine of contemporary cosmology:

- All aspects of visible physical world can be explained by the activities of four subatomic particles and four basic forces, which are the constituents of atoms. When physicists analyze the constitution of subatomic particles and forces, they enter into the realm of cosmology including standard model, Higgs mechanism, quantum gravity and mysterious entities. These deeper levels are mathematical models which cannot be verified experimentally; hence, they cannot be said to have existence in actual sense. It has been explained earlier that the standard model particles are verified only by indirect experimental methods of large hadron collider (LHC) and its efficacy is doubtful.
- The difficult puzzle is with regard to the earliest stage termed as *quantum gravity* in which matter and energy were combined as plasma. It represents the state of baby universe up to the stage of inflation happened at 10^{-35} second after big bang; in this period the diameter of universe grew from zero to about 10^{-17} centimeter only. This topic involves the search for an explanation of big bang including the question as to what existed before the origin. In this context, cosmologists have formulated a plethora of models about the prehistory of quantum gravity such as superstrings, big bang, membranes, multiverse, dark matter and dark energy. This area appears like scientific fiction.
- Through common sense we compare the original stage of our universe to a seed which grows to become a tree with multitude of branches and leaves; the seed is a unifying principle with maximum symmetry. But the mysterious models of quantum cosmology advanced **pluralism** about the most fundamental aspects of world – it is against the principle of symmetry and unification.
- The stages of quantum cosmology must be viewed in evolutionary and historical perspective. So we can say that the particles of standard model have evolved to become the subatomic particles

and forces of level II. **But theoretical physicists strongly propagate the wrong idea that the cosmological entities of levels III, II and I are the inner components of matter.** As a specific example, the much publicized string theory does not explain big bang and also subsequent events like inflation and expansion in the evolutionary history of universe. Hence, physicists do not have conclusive knowledge about the structure of matter or physical world.

- ❖ The position of scientific realism about the existence of fundamental particles and mysterious entities is problematic because such entities are mere inferences to the best explanation - it leads to the philosophical issue of *skepticism*. Here theoretical physicists are obsessed with scientism and realism.
- ❖ The various models of mysterious entities are not mutually supporting. For example, the description of gravitational waves, dark matter and dark energy does not connect properly with superstring-membrane theories. *The bizarre and imaginary models have not succeeded in solving the cosmological riddle.*
- ❖ Often the desire for prestige, power and money would drive theoretical physicists to establish hugely expensive experiments – mainly LIGO, LHC, neutrino research and satellite missions. These experimental pursuits have adverse impact on the goal of economic justice since the scarce resources are diverted for the mostly wasteful research establishments. Also, there are worrisome implications on environmental matters. It can be emphasized that the experiments related to mysterious entities of quantum cosmology must be assessed using the norms of ethics.

6.2 Solution to the Dilemma about Matter

The above discussion would boil down to the conclusion that proponents of theoretical physics are unable to show the stuff or building material of physical world. The existence of matter remains to be a puzzle in theoretical physics. As an implication, since the cosmological theory of matter is based on the theory of reality called *materialism*, it leads to severe philosophical dilemmas also.

In the wake of the above confusion, there is a general opinion among scientists as well as ordinary people that the very early stage of our physical world consisted of energy only. This view is the reflection of the religious idea that God created the physical world; here energy is treated as the manifestation of God. As an implication, the proponents regard that matter originated from energy; it is interpreted that **matter is an illusion.** Two types of interpretation are advanced for this epigram.

- Since the fundamental entities are mathematical constructions involving ten or more spatial dimensions, they have only instrumental value for explaining the historical development of physical world.
- The stage of grand unified theory (GUT) – including Inflation and vacuum at 10^{-35} second after big bang -- is treated as the source for the emergence of subatomic particles like proton, electron and neutron as well as the four basic forces. Since the effect of gravitational force belonging to matter was too small to be detected at that time, physicists hold that GUT stage consisted of energy only. Then they subscribed to the erroneous idea that material particles originated from unified energy. To strengthen this view, they treat particle as a packet of energy in tune with the equation $e = mc^2$.

The best selling popular science books of Fritjof Capra, Stephen Hawking, Paul Davies, Brian Green and John Gribbin have upheld scientific realism while describing the very early stages of universe. We may consider here only the writings of first and second authors for a critical treatment of the above curious idea about matter.

Matter Is Not an Illusion

Fritjof Capra, in his book *The Tao of Physics* (1982), provides a mystical interpretation for the notion of energy transforming into material particles. [# 10].

Capra tries to link the ontological issue of theoretical physics to mysticism, found in Indian philosophy of Vedanta as well as Chinese doctrine of Taoism. Quantum field theory (QFT) holds that the three standard forces were unified at the time of 10^{-35} second after big bang. We may comment that here QFT fails to consider the symmetry breaking process because it does not consider the presence of gravitational force which was extremely small without any considerable effect. In this situation, according to QFT, the tiny universe experienced sudden inflation for unknown reasons whereby its size increased by 10^{17} times to reach the diameter of one centimeter.

The mysterious event of inflation resulted in the GUT stage in which the universe consisted entirely of energy. Fritjof Capra interprets this energy field as *void or nothingness*, borrowing the metaphor from eastern mysticism. He adds that the void is a creative field of energy that caused the emergence of material particles. By a flight of spiritual imagination, Capra finds the similarity of void with the Indian notion of reality called Brahman. Here we can see an effort to explain the origin of physical world by resorting to the mystical view of cosmic reality. In other words, Capra unwittingly agrees to the spiritual approach of linking God with the physical world. It involves the process version of metaphysical realism that God and matter are real processes. We can recollect the critical philosophy of Immanuel Kant for showing the fallacy of Capra's approach. Especially, Kant has established the inappropriateness of metaphysical realism for connecting religious knowledge and scientific knowledge. Moreover, the process view is patently incapable of talking about the existence of void, Brahman and matter – content view only can deal with existence.

As an additional point of criticism, we can note that Fritjof Capra wrote his book in the heydays of quantum field theory which ignored the presence of matter or gravity. We have explained that the dual principle of matter and energy is applicable to GUT stage also, though the effect of matter was not manifested at that stage. The later advancements of new doctrines like quantum gravity and Higgs mechanism, which are placed before and after respectively of GUT, forcefully refutes Capra's view that matter is an illusion. It may be added that Fritjof Capra adopted process view with fallacious interpretation of body-mind dualism. Also, we can admit that Cartesian dualism cannot be overcome by process view. As an implication, the dualism of matter and energy also cannot be removed by process view. We must deliberate upon the existence of the opposites – matter and energy – by resorting to our System Philosophy of physical world.

In recent decades the exponents of **Vedanta** – the main religious doctrine of Hinduism – have been influenced by the best-selling books of Fritjof Capra. Accordingly, they are eager to advance the arguments to show the fusion of Vedanta and quantum cosmology. For that purpose they consider the GUT stage of our universe, which is metaphorically called as vacuum, void or nothingness. According to physicists, the GUT phase appeared as a highly creative form of energy from which the material particles emerged in accordance with Einstein's equation $e = mc^2$. Hence, the proponents of Vedanta interpret that the creative energy of GUT is the same as the cosmic consciousness called Brahman. In this manner, Brahman is regarded as the unifying principle behind the diverse particles of matter constituting the physical and biological world. We may call this mystic interpretation as Vedantic Cosmology. Its basic theory is that God is the source of the primordial energy of universe, which subsequently caused the formation of the material world. [# 11]

The foregoing interpretation seeks to connect the religious reality to the scientific phenomenon of GUT. Obviously, it is similar to the **mysticism** adopted by Fritjof Capra. However, we have already deliberated upon the serious drawbacks of mysticism as well as metaphysical realism. It may be reiterated here that religion and science are two separate levels of knowledge generated by our mind; these two forms of knowledge have different justifications in accordance with the tenets of System Philosophy. Hence, it is not possible coherently to explain scientific phenomena on the basis of religious doctrines. [*]

Next we may consider the concerned points from the famous books *A Brief History of Time* (1995) and *The Grand Design* (2011) by Stephen Hawking. [# 12].

In the first book, Hawking proposes the idea that the material world originated from the stage of quantum gravity in which the net energy was zero. Stephen Hawking expresses great confusion when he says that our physical world originated from the stage of *zero net energy*. We can explain this fact saying that gravity has negative energy and standard forces have positive energy, which cancel out due to the symmetry existing in the stage of quantum gravity.

Hawking in the second book explains that the fundamental theories of cosmology are justified by the doctrine of *model dependent realism*, which we can recognize as a synonymous term for scientific realism. In this way, the various mathematical models such as black holes, superstrings, membranes and multiverse are assumed to exist really as the fundamental aspects of physical world. It can be noted that the details of the mysterious entities given in the said book abounds in imagination and exaggeration, much beyond our objective sensibility.

The speculations about the origin of matter and physical world are aided by mathematical models, but it appears as a deviation from the familiar area of scientific method. We feel uncomfortable to think that cosmologists are propagating myths about the basic factors of physical world. There is a need to develop the philosophy of cosmology in order to escape from this vicious circle. **It is reasonable to point out that the origin of universe is an event which scientists cannot explain.** We have to enquire why cosmologists avoid philosophy and take refuge in illusory and complex experiments upholding weird theories. This scientific subject altogether is different from the religious account of the origin of universe as given in Bible and other Sacred Texts. The contrasting premises of science and religion with respect to the cosmological question will have to be kept in mind.

For a satisfactory account of quantum cosmology, the next step is to enquire about the nature of this knowledge. That is, the epistemological analysis must be undertaken to see whether the mysterious entities of the so called physical reality have existence. In this context we need conclusive answer to the deepest dilemma whether the role of God must be invoked for the ultimate explanation of physical world. In order to tide over this dilemma Stephen Hawking, Paul Davies and many other illustrious scientists have resorted to the notions of anthropic principle, fine tuning and Intelligent Design; that is going beyond the realm of science. This point will be explained in next section.

Coming to the puzzle of matter, we must consider it against the background of the mysterious entities. We have already explained the epistemological status of quantum cosmology establishing that the TyHDTI scheme cannot be applied there convincingly. In order to clear the philosophical mess, the only alternative is to accept the **system model** about the existence of physical world. Accordingly, we must accept the innovative idea that quantum gravity and its earlier stages are *fictitious descriptions of the physical reality* which actually is a system of matter and energy [*]. In this line of thought, matter and energy have complementary and constructive existence represented by X-axis and Y-axis respectively of coordinate system. Obviously, it amounts to the *content view* about the origin and development of physical world. The epigram that matter is an illusion is refuted here through the tenets of System Philosophy.

Regarding the existence of physical world as originated through big bang, matter and energy are the opposite forces which jointly produce the levels namely, quantum gravity, standard model, entities of quantum mechanics and the macro world of atoms as well as higher substances. We may note here that the matter-energy system can be formally called as space-time system where space has three dimensions and time has one dimension of straight line.

The True Meaning of $e = mc^2$

This is the right place to give the philosophical interpretation of the famous equation $e = mc^2$ discovered by Albert Einstein. As explained in section 1.3 of first chapter, this equation is based on the particle-wave duality (or space-time relativity) of subatomic phenomena. Specifically, it articulates the law of converting matter into energy and vice versa. This inter-convertibility has remained a puzzle to all great scientists including Einstein. However, we admit that nuclear energy is the outcome of the conversion of matter into energy. **But, is it really true to say that matter can be produced out of energy?**

I have already suggested that the true meaning of particle-wave duality is the following: *Every particle has inherent energy*. It means that every subatomic entity exists as a whole of material aspect and wave (energy) aspect. In other words, a particle is a combination of matter and energy.

But, as we have shown above, matter and energy are not really existing and independent entities. Rather, these are opposite entities which exist jointly as a system according to X-Y model. Then the **conversion of matter into energy and vice versa** must be interpreted in an ingenious manner as following. [*].

Consider two particles A and B represented by the points (x_1, y_1) and (x_2, y_2) respectively in the first quadrant of X-Y coordinate system. These particles are systems because of the general relation $y = xc^2$, where x stands for mass and y stands for energy. Let x_2 is higher than x_1 and let y_2 is higher than y_1. That is, particle B has higher mass and energy as compared to A. When particle B is converted into particle A through radioactive decay, there is a loss of mass as equivalent of $x_2 - x_1$; simultaneously energy equivalent to $y_2 - y_1$ is released. The point to be stressed here is that the conversion of matter into energy must be interpreted using the system model of different kinds of subatomic particles.

Solution to the Puzzle of Matter

Now we are concerned with the common sense question: **what is matter?** Here the defining property of matter is extension. So matter is contrasted with energy, which has wave nature. But subatomic phenomena have particle-wave duality so that matter and energy are inter-convertible; this makes the definition of matter ambiguous. We want to address the problem of existence of matter and energy. Since the spectrum of physical phenomena can be divided into many levels, we must consider the corresponding shades in the concept of matter. There are two separate notions of matter to be followed here as following.

Firstly, let us consider the notion of matter as applicable to the levels I, II and III including quantum gravity and standard model. In these cosmological stages, the entities are combinations of matter and energy in an inseparable way. For example, a quark is a combination of matter and energy, where matter is a quantified form of energy. In the case of such cosmological entities, there are special issues connected to the conception of space and time. Additionally, the notion of gravity is not explicitly considered here, not withstanding our suggestion that the Higgs field can explain the phenomena of gravity. *Due to these problems, the concept of matter does not have absolute and practical relevance in the said cosmological stages.*

Secondly, we have to take up the levels of micro subatomic world (IV) and macro world (V). The four subatomic particles and four basic forces adequately explain the formation of atoms and higher objects in the visible world. All these particles and forces have the curious property of particle-wave duality. For example, electron can be seen as a particle in some situations like the picture tube of a TV, while it is a wave in the context of some other experiment. The case of a basic force is also similar. But *we conventionally accept the distinction between particle and wave in visible micro and macro world.* That is why we have separately named the four subatomic particles and four basic forces – it is through the application of the rules of thought originally proposed by Aristotle. [# 13].

In the visible level, the basic unit of physical world is atom formed by the planetary structure of the three particles namely, proton, neutron and electron. There are about 110 different kinds of atom on account of the variations in planetary model. The three dimensional extension – length, breadth and width – of an object is originally due to the planetary model of constituent atoms. Hence, atoms are like the bricks of a building. The visible level of physical world consists of extended substances and energy radiations. Note that the energy radiations – for example, light rays and x-rays – are caused by the interactions of basic particles and forces inside the atoms.

Considering the above features of subatomic level and visible level we adopt the **practical notion of matter** [*]. The property of extension of matter is basically due to the planetary structure of atoms; such atoms constitute the visible structure of physical world. It implies that the constituent subatomic particles themselves are not treated as matter; they are just material particles having particle-wave duality. In other words we cannot say that matter exists as a subatomic particle because the absolute notion of particle is not relevant here. *Accordingly, the word matter refers to the individual atoms and the higher substances formed by them.* In this situation, we observe matter and energy separately because of our conventional rules of thought.

Now we can agree that matter and energy are generic terms which have variable meanings as per the level of physical world concerned. As per *System Philosophy*, matter and energy are opposite parts of phenomenal systems which constitute the particular level of physical world. The system model of universe effectively unifies the levels of physical world. However, the term matter does not have clear meaning in the case of cosmological entities including standard model [*]. ***In our ordinary usage, matter refers to the atoms and higher substances.*** This matter is constituted by subatomic particles and basic forces. On the basis of the layered view of physical world, we reach at the solution for the puzzle of matter.

The foregoing conclusion about matter can effectively deal with the dichotomy in **cause-effect** relation also. The notion of cause-effect is deterministically defined in the case of material bodies of visible world. But subatomic particles and lower levels do not follow this kind of causality on account of particle-wave duality and uncertainty principle. In other words, the causality of quantum mechanics involves the element of chance; hence it is different from that of classical science. This dichotomy is now reconciled on the basis of the discussion of previous paragraphs.

6.3 Frequently Asked Question about the Origin of Universe

We may now consider the frequently asked question about the origin of universe. This is the basic cosmological question. Here the term *universe* denotes the entire expanse including galaxies, stars, planets and other areas, which is known to us by our intellectual faculties. This universe consists of various levels, mainly inanimate world, biological world and social systems that can be experienced through our sense organs. Obviously we exclude the metaphysical concepts like God, gods, angels and evils from our purview. In this context we may restate our problem as: **how did the universe originate and evolve?**

In the history of human thought, the above question is sought to be answered in two alternate ways as the *religious view* and the *scientific view*. According the religious view, this universe is a creation of metaphysical agencies, conveniently represented by the notion of God. Since the idea of divine creation cannot be linked to the physical concept of time, the religious view leads to many inconsistencies when we talk about *origin* and *evolution* of universe. Accordingly, the concerned theories about God and world - Idealism or Theism, Deism, Intelligent design argument (IDA), mysticism and Vedantic cosmology – need not be considered here. In this situation, we have to shift our focus exclusively to the scientific view.

As per scientific view, the things of universe are observed in physical terms involving the basic concepts of space and time. In this manner, inanimate things are bodies made of matter and energy – this is the basic level of *physical world*. Here matter and energy are interconvertible; hence we can hold that the concept of matter includes energy. Additionally, science treats biological world and social systems also as physical levels – life and mind are said to be the algorithms of the material activities of body. Then the basic cosmological question can be restated as: **how did matter originate and evolve?**

Science seeks to answer this problem on the basis of **materialism**, which is the theory of reality that the first cause of universe is a being or substance called matter; it is according to the content view of knowledge. We know that classical science upholds this view envisaging that matter is made of tiny and homogeneous particles called atoms. In contrast, modern science is an edifice built upon the pillars of quantum physics and quantum cosmology; it adopts the process version of materialism, specifically known as Physical Process Reality. We have to take into account the salient features and philosophical issues of classical and modern science, as covered in the previous chapters, in order to deliberate upon the basic cosmological question.

To cut the long story short, the main and concerned findings of quantum cosmology are restated now. Matter (physical world) originated by the event called big bang. Latest research has explained that the big bang is not an explosion, but it denotes the *beginning from zero size*. After big bang, the diameter of physical world grew to 60 km in 10^{-4} second, when the elementary particles like quarks and leptons and bosons emerged; to 6,00,000 km (6×10^5 km) in 1 second causing the formation of subatomic particles and basic forces. Around 9.10 billion years after big bang, earth was formed; subsequently, living beings emerged through evolution. And at present, that is 13.7 billion years after big bang, the diameter of physical world is 10^{23} km and it contains 10^{50} tons of matter in the form of atoms, molecule and compounds.

On the basis of these facts, our *basic cosmological question* is to be split into **three problems** as following:

- ❖ What did exist before the big bang?
- ❖ How did various types of particles and forces originate so as to form the galaxies, stars and planets?
- ❖ What is the explanation about the emergence of life upon earth and also the evolution of a hierarchy of species?

Physicists have been trying to find the more elementary components of subatomic particles and basic forces for explaining the origin of matter. Their argument is that the components of a thing must occur before the existence of the thing. Hence, the more elementary the component of matter, the earlier is its origin. Prompted by this idea, theoretical physicists have used experiments and mathematical models to propose that standard model particles, quantum gravity, strings, membranes, multiverse, dark matter and dark energy are the earlier stages of matter in the backward direction of time. Recently physicists have interpreted big bang as the event of *compactification* of entities that have ten or more dimensions; where by the extra dimensions are hidden to result in the four dimensions of space-time.

The above theory has validity only if the cosmological entities have real existence. For example, for *compactification* to happen, strings and membranes must actually exist. How can one say that the concerned mathematical models represent real aspects of the early stage of universe? This is properly the problem of justification. Challenged by the possibility of skepticism, physicists adopted the position of **scientific realism,** which attributes real existence to theoretical entities of cosmology from practical point of view.

However, there are five serious **drawbacks** in these scientific proposals of recent decades.

1) The search for the most elementary form of matter is elusive, because physicists are increasingly resorting to mathematical models associated with the notions of symmetry and group theory.
2) The experimental evidence is lacking, or at least ambiguous, in the case of the said mathematical models.
3) There is the danger of *pluralism* in the case of elementary components of matter and this situation goes against the idea of physical reality or stuff of universe.
4) From philosophical angle, we can say that the *scientific realism* is a wrong assumption. The so called elementary components of matter as well as the event of big bang are constructions of scientific mind with pragmatic motives. We are not entitled to propose the existence for such entities.
5) Scientists have failed in explaining the evolution of matter to more complicated forms and also the emergence of life and mind

causing various kinds of biological cells and organisms. The biological world abounds in nonphysical aspects of creativity and purpose, which are outside the scope of science.

Realism promotes the quest of scientists for finding more elementary levels of matter, even if through mathematical models, without any advancement of real knowledge. We may note that scientific realism is bed rock of materialism, strongly promoting scientism and atheism. Then the conflict between science and religion emerges as an intractable problem. It is a moot point whether the hugely expensive experiments of LHC, neutron research, study of gravitational waves, LIGO and so on, bring significant benefit to human society, other than serving the professional and material interests of concerned scientists. When half of world's population is poor and suffering, the principles of ethics are to be observed in scientific research.

In a summary way, we can say that the three problems mentioned above remain unanswered, in spite of spending colossal amount of money and effort for cosmological research. Scientists have reached the dead end. The silver lining in this cloud comes from the principles of System Philosophy of Science developed in this book. **My answer to the basic cosmological question** is given concisely as following.

First. It is asserted that the theories like big bang, inflation and expansion of physical world and Higgs mechanism are meant only *for scientific way* of interpreting the history of physical world. We can say scientifically that the physical world originated about 13.7 billion years ago through the event called big bang, which denotes the beginning of *compactification* of past universe having ten or more dimensions.

Second. The past universe (parental universe) is a system of dark matter and dark energy, which are denoted by DM and DE respectively. **This system has existence** since it is represented by the X-Y model of coordinate geometry. Then big bang is the origin ($x = 0$, $y = 0$) of the system model. The opposite entities DM and DE can undergo the process of compactification throughout the history of present universe;

it accounts for the increase of matter from zero to 10^{50} tons. We can compare our present universe as a plant, which grows by drawing resources from land – here land includes water *and* atmosphere also. Then big bang is similar to the stage of seed, while past universe (DM-DE system) serves as the land. This analogy helps us to answer the question as to what existed before the big bang.

Third. In order to show the existence of past and present universe in a historical manner, we can use similar X-Y model. Then, the past universe appears in third quadrant while the present universe (physical world) belongs to the first quadrant. The X-Y coordinate system is called the **System Model of physical world** according to the content view; then the point (0, 0) is the singularity point of big bang. The hierarchical levels of our physical world – subatomic phenomena, galaxies, stars and planets – are produced by the dialectical and productive relation between matter and energy.

Fourth. The above *System Model of physical world* answers the questions about the origin of our physical world, since it can be transformed into another X-Y model, where X denotes space and Y denotes time. It is emphasized that *origin* and *evolution* are physical notions, which are to be conceived in the framework of space and time.

When we ordinarily say that our physical world originated through big bang, we assume that big bang actually happened. It is the position of scientific realism which must be refuted. For that purpose, we hold that the above *system model of physical world* is a theory; it describes the past universe as DM-DE system, while present world is a system of matter and energy. Then the TyHDTI scheme is applied to get the inference that physical universe exists. In other words, **existence is an inference** produced by the cognitive mind of a person; the realism is avoided through this way. Also, the *system model* effectively removes the problem of pluralism.

On further deliberation, we must admit that there is *nonphysical* purpose and creativity in nature. This feature is visible in inanimate world (including galaxies, stars and planets) and more clearly in the phenomena of biological world. It will be shown in due course that nature is a *matter-consciousness system*. We do not require *the intelligent design argument* for explaining the *nonphysical aspects*. Without assuming realism, we can show that the scientific view of physical world is the materialist reduction of matter-consciousness system. More specifically, the DM-DE system or matter-energy system is the **physical reduction** of matter-consciousness system of inanimate world. [*]

When we say that physical world (matter-energy system) is the construction of our scientific mind, we mean that it is phenomenal. Since justification is a part of secular epistemology, it has scientific connotation. In the layered view of universe, the topic of justification is concerned with the separate and hierarchical existence of various levels of systems which are broadly grouped into inanimate, biological and mental worlds. Later we will develop the *System Philosophy of Ultimate Reality* showing that the Ultimate Reality or *paramporul* is depicted by X-Y coordinate system of body and consciousness.

As already explained, the system model of physical world, connecting dark matter and dark energy through X and Y coordinates, has a clear advantage. Since space and time are physical concepts, we can directly explain the formation of the subatomic particle and forces (the world of quantum mechanics) as well as higher substances using this method. It is phenomenal knowledge. In this manner the *System Philosophy of Science* is capable of addressing the issues about the epistemology and ontology of physical universe. However, there is a mystery in the purposeful development of nature; it is to be explained holding that nature is a system of body and consciousness as mentioned above. The notion of time is not applicable at the level of reality. Hence, we cannot bridge the conceptual gap between reality and phenomenon.

6.4 Do We Need Intelligent Design Argument?

Now we will deliberate upon the serious confusion prevalent in a section of eminent scientists and authors of quantum cosmology. They are troubled by the doubts about the explanatory power of the mathematical models of mysterious entities. The notion of multiverse, membrane and standard model are presented in a factual and scientific manner; it tries to answer the *how* question about the origin of universe. But through deeper thought, the concerned authors have recognized that the cosmological models do not explain the *why* question with regard to the cosmic structure and evolution; it is elaborated below.

According to cosmologists and scientists, the existence of physical world and its evolution depended on a number of cosmic coincidences. So far, they could identify around 200 cosmological constants which serve to set the numerical values of the parameters of subatomic phenomena. A few examples may be given now.

- The values of the mass and electric charge of four subatomic particles as well as the difference in the strengths of the four basic forces are crucial for the formation of various atoms and higher substances.
- The relative effects of antigravity and gravity is responsible for the expansion of universe and the distribution of matter in galaxies. If the strength of gravity in comparison to other forces was slightly higher, then the entire matter of universe would have clumped together.
- Additionally, there are certain parameters regarding the conditions of earth, including atmosphere and climate, which make earth as a unique planet suitable for life. The presence of oxygen in the atmosphere, abundant formation of water, soil structure and a lot of other factors contributed to the formation of first cell with life around 3.5 billion years ago.

In the span of 13.7 billion years after big bang, the first 10.2 billion years belonged exclusively to inanimate world; it evolved from minute particles to galaxies, stars and planets. But suddenly life emerged in a cell formed upon earth, a tiny spot in the infinitely large expanse of astronomical bodies. The next 3.5 billion years saw the evolution of life into millions of species of organisms. It culminated in the advent of human species designated as homo sapiens around 40,000 years ago. The saga of the evolution of inanimate and biological worlds displays the process in which some simple things are combined to form complex things and the repetition of such processes in numerous ways. This process of increasing complexity has resulted in the hierarchy of inanimate and living things. It can be treated as the evidence for *creativity, freedom and purpose* in universe. We cannot say that the universe is an arena of random events like the play of dice. The recognition of purpose in nature is the key to deliberate upon the *why* question related to cosmology.

The calculations about the cosmological constants would indicate that the physical world was subjected to *fine tuning* in order to make it fit for the emergence of life upon earth and subsequent development of living organisms. Fine tuning can be treated as the empirical evidence for the purpose in nature. In this context, a group of cosmologists argue that the fine tuning happened for the existence of human beings. This statement is popularly known as *anthropic principle*. The proponents of anthropic principle emphasize that the fine tuning of inanimate astronomical world is not solely attributable to chance; instead, it suggests the existence of an Intelligent Designer with creativity and purpose, as the original cause of physical world. [# 14].

We can scientifically observe the creative and purposeful aspects of world. Though such properties are certainly nonphysical, it is possible to define them in physical terms. For example, the diversity of physical substances such as gold, water and air as well as the cosmological constants are clear evidences of creativity and purpose in the universe. Taking into account such empirical facts, the religiously oriented scientists have proposed the idea of Intelligent Designer, which imply two attributes as following:

- Intelligent Designer is an infinite power with creativity, freedom and purpose.
- Intelligent Designer contains the laws for the design of universe.

It is important to note here that the Intelligent Designer is the empirical version of religious God. Accordingly, the belief in Intelligent Designer is an Inference to the Best Explanation (IBE), which is associated with scientific realism. Moreover this notion suffers from the problem of induction and also skepticism – these are sound reasons for rejecting it. This lacuna of Intelligent Design Argument would be further discussed in the companion book *Life and Mind*, in the context of Darwin's theory of biological evolution.

We may temporarily shift our deliberation to *philosophy of religion*. The belief in God is traditionally used for answering the "why" question related to the creativity and purpose in universe. The philosophical issues related to the concept of God also will be presented in the next Book. However, the main points may be given below. [# 15][*].

1. God and allied concepts like soul, angel, heaven and hell are formed by our mystic mind, which is responsible for religious worship.
2. Mystic mind is complementary to scientific mind. Hence, the existence of God cannot be proved using scientific arguments. There are additional problems in connecting God to physical world. The so called evidences about the nonphysical aspects of world cannot serve to justify the mystic notion of God.
3. The existence of God as well as the issue of God's existence must be conceived in the framework of mystic mind (religious mind) only. These ideas are expressed in metaphoric or nonscientific language.
4. Even within the premise of religious mind, there are many serious issues to be settled about the attributes of God. For example, when a believer says that "God is infinite" and "God

is love", these sentences need to be interpreted metaphorically in tune with the objectives of religion.

Since Intelligent Designer is a jump of imagination from finite evidences to the realm of infinity, there is no coherent meaning for that concept. Unfortunately a section of scientists and theologians have attempted to promote the idea of Intelligent Designer, without knowing its epistemological drawbacks. **In a summary way, we can assert that there is no justification for their empirical inference about an infinite being.**

Having dismissed the intelligent design argument as a fallacy, we get a clear perspective to interpret Stephen Hawking's references to God [# 16][*].

When Hawking uses such expressions as *mind of God* and *benevolent creator*, he does not intend to bring the religious idea that God created the physical world and its laws. We can interpret that it was a metaphoric way of speaking about the fine tuning as well as anthropic principle. The implication is that the ultimate cause of physical laws and the evolutionary development is beyond the scope of science. As explained above, Stephen Hawking and other concerned cosmologists invoked the empirical idea about intelligent designer as a short cut to tide over the difficulties in explaining purpose in universe. However, they adhere to the materialist path, specifically the physical process view, for the factual doctrines of quantum cosmology.

In the final part of previous section, we have stated that the physical world exists like a factory designated as matter-energy system. But it is necessary to ask: what is the source of the purpose and creativity visible in the evolution of physical world and emergence of living organisms? Here we are considering the nonphysical aspects, which are required to explain physical world philosophically. This vital question will be answered thoroughly in the third book *Discovery of Reality*. For that purpose, we have to adopt the idea that reality is a matter-consciousness system. But physical world has phenomenal existence only. The frequently asked question about the origin of universe, taken in physical sense, has already been addressed. So we may anticipate that the distinction between phenomenon and reality is crucial for completing our discussion about natural things. In this way we can finally solve the conundrum of God and Intelligent Designer in the context of science [*].

NOTES of Chapter 6

1   Davies (1995) and Green (2005) have discussed the *arrow of time* in great detail. However, by linking *arrow of time* with cause-effect, we can note that time always move from left to right. (*this is my original idea)

2   The most comprehensive treatment of space and time can be found in the following books -- Davies (1995), *About Time – Einstein's Unfinished Revolution;* Green (2005), *The Fabric of the Cosmos.*

3   Kant, Immanuel (2003), *Critique of Pure Reason*, translated by J. M.D. Meiklejohn (Dover Publications, New York, 2003), Transcendental Dialectic, Book II, Chap. III, Section IV, pages 331-336;

4   The critical philosophy of Immanuel Kant is adapted here for explaining that the fundamental entities of quantum cosmology are mere predicates. (*this is my original idea).

5   The layered perspective is based on the table of the structure of physical world, given in Table 1 of chapter 3. (*this is my original idea).

6   Davies (1995), pages 24, 132, 185-186. The theological perspective of creation is a consequence of the idea that *time has an absolute beginning at big bang*. Accordingly, our physical level is a creation of God. As per this belief, God is outside of

time; this causes a conundrum. We can quote from page 24 of Davies (1995): "the core of the debate is the daunting problem of how to build a bridge between God's presumed eternity on the one hand and the manifest temporality of the physical universe on the other. Can a god who is completely atemporal logically relate in any way at all to a changing world, to human time? Surely it is impossible to exist *both* within and outside of time? After centuries of bitter debate there is still no consensus among theologians about the solution to this profound conundrum".

7   Hawking (2011), *The Grand Design*, Bantam Books, 2011 edition, page 130. In section 3.3.3 of chapter 3, we have explained that the phenomenon of space-time curvature is valid only in the context of total universe, which is the sum of past and present universes. The presence of *dark matter* and *dark energy* also must be taken into account for explaining the curvature of space-time. (*this is my original idea).

8   Green (2005), *The Fabric of the Cosmos*, page 419.

9   Out of the many reference materials available about Gravitational Waves and LIGO experiment, the following two sources may be specially mentioned:

a)   Green (2005), *The Fabric of the Cosmos*.
b)   The article, entitled *The Discovery of Gravitational Waves and the Future of Religion and Society*, by Job Kozhamthadam published in June 2017 issue of *Omega– Indian Journal of Science and Religion* (Institute of Science and Religion, Aluva, Kerala)

10   Capra, Fritjof (1992), *Tao of Physics*, pages 27, 65, 77-93 and 220-272. Capra did not understand the distinction between phenomenon and reality. He wrongly proposes that the real aspects of body and mind are the same as the empirical

processes of physical and mental phenomena. But Descartes had conceived body-mind dualism from metaphysical perspective.

11   Two books dealing with Vedantic cosmology are:

a) Jitatmananda, Swami (2006), *Modern Physics and Vedanta* (Bharatiya Vidya Bhavan, Mumbai 2006)
b) Panda N. C. (1999), *Maya in Physics* (Motilal Banarsidas, Delhi, reprint 1999)

12   Hawking (1995), page 136; Hawking (2011), pages 16, 57-68, 216-217

13   Aristotle's rules of thought has been introduced in section 5.2 and it will be explained further in the third book *Discovery of Reality* also.

14   The notions of cosmological constant and antigravity are explained in: Davies (2007) pages 54-61 and Green (2005) pages 270-286. Stephen W Hawking and Paul Davies have talked about intelligent design argument in their famous books.

15   The contrast between scientific mind and mystic mind is my original idea. And, the concerned reference books about *philosophy of mind* are shown in the selected bibliography.

16 Here the references are: Hawking (1995) pages 149 and 185; Hawking (2011) pages 38-42 and 207-210.

Chapter 7

Further Vision of System Philosophy

****

7.1 Life and Mind

- Life and Evolution
- System Philosophy of Mind

- System Philosophy of God and Evil
- Science- Religion Synthesis

7.2 Discovery of Reality

o A Guide to the Levels of Knowledge
o Philosophy and Its Main Divisions
o World and Reality
o Existence of Seven Life Systems
o Comprehensive View about Truth
o Glimpse of the Path Ahead

Author's main original ideas are marked by [*].

The mark [#] gives the number of note at the end.

****

Since we have discussed the various aspects of physical world in the foregoing chapters of the present book, *Origin of Universe*, it is now important to extend our analysis to the living world also. Thus we want to develop the philosophy of science pertaining to the basic aspects of biology and psychology. Additionally it is necessary study the features of our Social Systems, mainly the religious system. All these intellectual efforts are aimed to formulate an integrative theory of reality and truth. For the expediency of philosophical treatment, we have divided this range of subjects into two companion books titled *Life and Mind* and *Discovery of Reality* respectively, within the framework of System Philosophy. The following sections provide an overview of the topics covered in those books, in order to complete the content of the present volume.

7.1 Summary of the Book : *Life and Mind*

The main contents of chapters are presented concisely in the ensuing sections.

Life and Evolution

Scientists speculate that life originated upon earth consequent to the evolution of material things during the earlier stage of this planet. But we feel that the emergence of life is an epoch that needs explanation from scientific as well as philosophical point of view. The secret of life will be revealed only if we answer the following questions: What is life? How do we get knowledge about life? What is the theory of modern biology? How did life originate in the earliest form of living beings called bacteria? What is the relation between life and material body? How can we explain the biological evolution leading to the diversity and hierarchy of living beings?

The fundamental ideas of modern biology – with the main branches called cell biology, molecular biology and genetics – are presented here concisely for a meaningful discussion about life and evolution. By criticizing the concerned scientific theories we will proceed to develop the epistemology of modern biology. We can then introduce the System Philosophy of Life and Evolution. This innovative philosophical framework will radically interpret the latest scientific achievements in biology and mark its limitations. [# 1].

The topics to be discussed are the following:

- ❖ The Puzzle of Life
- ❖ Modern Biology
- ❖ Criticism on Computer Model of Life
- ❖ System Philosophy of Life
- ❖ Theory of Evolution and its System Philosophy

The practical distinction between life and mind is necessary for deliberating upon the philosophical issues pertaining to biological world. The main characteristics of life are metabolism, growth, adaptation, reproduction and self-maintenance. When we consider the history of human thought, there are *five different worldviews* for defining and explaining life and related phenomena [*]. The enigma of life is primarily due to these worldviews pertaining to fact. As a result, religion and science proposed divergent concepts about life.

The modern biology aims to describe the functions of various parts of organism. The shift of focus from structure to function is achieved through a new framework called *machine-algorithm model* upholding the *physical process view* [*]. The interrelated activities of cellular parts as a whole can be treated as an *algorithm*. Accordingly, the term *life* refers to the totality of activities happening at various levels of physical body of organism. The key concepts are cell theory, cell specialization, the central dogma of molecular biology, DNA, gene, genetic code, genome, genetics and machine-algorithm model. This scientific view aggravates the puzzle of life.

It can be shown that the machine-algorithm model(computer model) of DNA and genetic code is a mechanical and physical way of describing the aspects of life. Further we explain the philosophical aspects -- epistemology and ontology -- about the concerned propositions [*]. Mainly, the existence of separate entities called DNA and genetic code cannot be proved, since they are predicates as per the philosophy of Immanuel Kant. Moreover, genetic code displays nonphysical aspects of creativity, freedom and purpose. In this milieu, there is no justification to the theory about DNA and genetic code.

System Philosophy of Life shows that DNA and genetic code exist together as a system formed by the opposite entities called *macromolecule* (mainly DNA and protein) and *information*. The phenomenon of life has non-dual existence in the form of a system. It is due to the influence of the Aristotelian rules of thought that we conceive macromolecule and genetic code as separate objects [*].

Next we discuss critically the physical process theory of evolution, specifically Darwinism and its new version namely neo-Darwinism. The four stages of biological evolution – *variation, struggle for life, inheritance and natural selection* – are examined to expose its drawbacks. *Then it is necessary to switch over to the principle that evolution must be perceived as a phenomenon under content view of knowledge - it leads us to the System Model of Biological World* [*].

Accordingly, every organism has physical and nonphysical parts, or alternatively matter and consciousness. It may be clarified that here consciousness is responsible for the purpose, creativity, design and planning observable in the stages of biological evolution. The system of macromolecule and information (X-Y system) has phenomenal existence and it functions as the cause of evolution. The mutation of interrelated genes of a species and natural selection altogether is a creative event which results in the formation of a higher species. This is the *System Model of Biological Evolution* under content view [*].

The formation of a new species from an ancestral species happens on account of four factors namely mutation, struggle for life, inheritance and natural selection. *System Philosophy of Biological Evolution* rejects the materialist approach, by holding that each of the four factors is a combination of matter and consciousness [*]. Further we can show that Darwin's theory is a physical reduction of the system model being proposed here. It would equip us to tackle the *problem of missing links* in an ingenious way. In this context, we can criticize the Materialism and Atheism of Richard Dawkins. The acrimonious debate between evolutionary scientists and creationists is the result of the ideological differences. [*]

System Philosophy of Mind

Higher organisms, especially human beings, have three levels of structure as physical body, life and mind. The term *biological body* is used to refer to the combination of physical body and life. Biological body consists of cells, tissues, and organs in a hierarchical order; all these parts have specific biological functions. But the special organs such as brain and sensory organs are together called *nervous system*, which additionally perform certain mental functions such as feeling, willing and thinking. The totality of mental activities is conventionally termed as **mind**. Accordingly, we hold that *mind is a higher phenomenon which exists over and above the biological processes of nervous system.* Here, we distinguish between the functions of life and mind.

In this context, we can treat **mind** as a factory which produces various mental phenomena like emotions, desires, thoughts and memories and at the same time as a being which controls various activities of biological body. We are most aware of our own mind, but are doubtful whether mind exists in lower beings like plants, birds and animals. The difference in the mental capabilities from person to person is highly baffling. At the same time, we often wonder how human species acquired the complex form of mind in the course of biological evolution. The religious interpretation about the phenomenon of mind adds to the mystery in this field of enquiry. We may tentatively accept that *philosophy of mind* aims to explain the deeper questions about the existence of human mind and its activities.

The foremost issue in *philosophy of mind* is the definition of mind since we have to take into account the related notions like body, soul, spirit and consciousness. There are great differences between science and religion while considering the question: what is mind? More specifically, the development of psychology as the scientific study of mind is at loggerheads with the religious approach. It is necessary to reconcile such conflicts by adopting a logical and secular point of view. We can anticipate that the system model of life and evolution presented earlier would show the path for dealing with the next level that is mind. This approach will lead to deeper insight about the ultimate reality behind the inanimate and living things of this universe.

In order to make headway, we would deliberate upon the traditional dilemmas *regarding mind, as existing in human being,* because these are relevant in the philosophy of science also. The latest advancements in scientific study about mind – various branches of neuroscience and psychology – constitute the ground from which we will start our discussion. Through philosophic review it can be shown that the three doctrines under *physical process view* – identity theory, behaviorism and computer model functionalism (CMF) - have serious drawbacks.

The central objective of the **System Philosophy of Mind** is to examine the doctrine of neuroscientists that mind exists above the *brain and other parts of nervous system (BNS)*. The scientific view does not recognize the nonphysical aspects of mind like creativity, purpose and freedom. It leads to the conflict with the religious (idealist) theory that mind exists as a metaphysical being, which is traditionally called as the mind-body dualism. [# 2].

Our main conclusions are given below. [*]

- ➢ By virtue of the system models, we establish that BNS is not a structure of physical organs. Adopting *content view*, human being is a system of matter and consciousness; this system has three levels of organizations, namely inanimate body, biological body and mind.
- ➢ Biological body and mind are separate, but mutually dependent, subsystems of human being that exists as a system. When these subsystems are treated as two levels of matter-consciousness, there is no dualism per se. Hence the age-old problem of *mind-body dualism* is solved.
- ➢ We can treat the mind as a factory with X-Y model, which produces three main classes of outputs namely *orderly neural*

networks, unconscious mind and *conscious mind* – these are interconnected. Each mental state is the combination of the said three main classes of outputs.

Matter and consciousness exist as complementary opposites constituting the reality of mind. The reality, as a production system, produces various levels of phenomenal mental states. The consciousness of a mental state is the phenomenal version of the real consciousness pertaining to the reality of mind. *Self-consciousness* is a conventional phrase denoting the *high level of consciousness of* human beings as compared to lower beings. [*]

System Philosophy of God and Evil

We have already introduced the idea that our cognitive mind consists of the dual parts namely *intellectual mind* and *mystic mind*. The former tries to understand things of world using sensory data and rational thought; it has two parts -- lower part is the *philosophic mind* and the upper part is called *scientific mind*. On the other hand, mystic mind produces the experiences and knowledge about religion as well as art. Then the method of philosophy, under the faculty of philosophic mind, can be qualified as *superscientific*.

In the precise way, *philosophy of religion* is defined as the epistemology of religious knowledge. On account of divergent worldviews, this subject has failed to suggest a method for unifying the various religions (forms of worship and theologies). These separate religions appear to have different sets of beliefs and activities of worship. *Theology* literally means discourse about God and it includes the rigorous interpretations of beliefs and practices for defending a particular religion or sect. In that situation, the philosophical examination of religious propositions must concentrate on the common or core ideas of various theologies, such as existence of God or gods, divine attributes, reason and faith, after life, souls, creation and problem of evil.[# 3].

The discipline called philosophy of religion has reached **a state of crisis** mainly due to the following reasons [*]:

- There is a conventional view that separate religions are different paths to reach God. Since descriptions of God under various religions are distinct, and often contradictory, it is against logic to say that they talk about the same God. How can it be that Christian God is identical to Hindu gods?
- The existence of God remains as an unsolved problem in the history of philosophy of religion, since scientific reason cannot show God's existence. What is the meaning of existence, if it is conceived by mystical mind?
- The social aspects of religion is conspicuously absent from the philosophy of religion as proposed so far. This issue is compounded by the question: Why are there visible evils in the organizational aspects of religions? Taking this fact into account, the definition of God needs to be modified drastically, so as to accommodate the notion of Evil also.

System Philosophy gives a radically new vision about the existence of *religious life System* (RLS), which is the totality of all religions as social systems, formed by the religious faculty of human mind. This fact is to be used for developing *system philosophy of religion* [*]. We reach at the conclusion that the existence of God is an inference confirmed by the mystical experiences of a worshipper. The rational and empirical doctrines of theology will be synthesized to get the system model of God and Evil, which can remove all issues in the field of religious knowledge. As the idea of existence is not part of the theory (theology), the problem of metaphysical realism is removed. Thus, the central argument of atheism can be firmly refuted [*].

Science- Religion Synthesis

So far we have treated the propositions of science and religion as two distinct forms of knowledge produced in our cognitive mind.

Now it is necessary to discuss the conflict between the assertions of science and religion. The outstanding problem is introduced by the author as following: Science denies the existence of supernatural beings or forces. It reduces the natural things, which are actually composed of body and mind, into forms of matter and energy. That is, science holds that everything in the nature is *physical*. It formulates cause-effect relations, known as *physical laws*, based on sensory or experimental evidences about the properties of physical things through our *scientific mind*. This is public knowledge in third person perspective.

On the other hand, religion presumes that supernatural beings or forces exist and they involve in the affairs of natural world. This is based on mystic experiences through revelation, emotions, ecstasy and meditation of religious leaders as well as ordinary believers. The mystic ideas are private and beyond sensory experience - it is the function of *religious mind*. [# 4] [*].

A coherent argument is developed for solving the issue. System Philosophy shows the synthesis of science and religion by treating them as two parallel systems of knowledge, which are two levels in terms of methodology, source, justification and truth. More specifically, the unification of science and religion is achieved because these are two kinds of knowledge formed by two separate faculties – scientific mind and mystic mind respectively -- of human mind.

It is a religious idea to believe that God, Evil, soul, heaven and hell, angels and so on exist as separate entities in supernatural realm. Through our intellectual faculty, we can show that the Ultimate Reality of universe is an X-Y coordinate system of body and consciousness. It causes the formation of a hierarchy of things including inanimate and living things, in accordance with the layered view of universe. The existence of human mind with various levels of faculties is explained in this view. This is the final justification of the science-religion synthesis proposed above.

7.2 Summary of the Book : *Discovery of Reality*

We are now focusing on the third book in the series, which deals with the fundamental aspects of world and phenomena. Its contents are essential to conduct the philosophical analysis as presented in the previous two volumes with regard to science, religion, social science and other forms of knowledge. Firstly, we have to familiarize with the basic features of knowledge so as to open the gateway to the enquiry about the integrative vision of reality.

A Guide to the Levels of Knowledge

As already mentioned, the principal aim of this treatise is to deliberate about the method, validity and truth of various branches of knowledge. The discussion runs through the following subsections :

- ❖ Definition of knowledge and its classification.
- ❖ Organic levels of propositions
- ❖ Approach to theory of knowledge

Here, the innovative idea is that the innumerable topics or disciplines can be finally brought to a 2x2 Table using the criteria of Content View, Process View, Rational part and Empirical part. [*]

It is the traditional practice to study the topic of knowledge by considering the level of individual propositions; but it has caused many philosophical issues. Hence the radical proposal of this chapter is that *the study of knowledge **must be shifted from the level of propositions to the level of methodological stages*** called Theory, Hypothesis, Deduction, Testing and Induction – these stages are abbreviated as Ty, H, D, T and I respectively. We may use the phrase *TyHDTI scheme* to denote the method of producing a particular discipline of knowledge. This way of analysis is efficient to apply to various kinds of disciplines [*].

It is instructive to hold that knowledge has hierarchical levels in a similar way as the levels in biological world such as cell, organ, organism and species. Aggregation of disciplines in successive stages will reach finally at the macro fields or kingdoms of knowledge called science, religion, art, philosophy and spiritual science. Salient features of these subjects are mentioned for the benefit of lay readers.

Philosophy aims to study the basic aspects of knowledge about world consisting of *observable* things and *unobservable* things. It has two main branches called ontology (theory of reality) and epistemology (theory of knowledge). The ontology deals with our notion of reality or first cause of universe. But epistemology is concerned with the essential aspects of knowledge, which are namely **methodology, source, justification and truth**. The philosophical questions about knowledge must be addressed from this point of view. [*].

For dealing with the deeper components of ontology and epistemology, we must have sound understanding about the subject matter of philosophy. A thematic survey of the competing doctrines of philosophy is given in the second chapter of this book. Its brief indication is given below.

Philosophy and its Main Divisions

Science, religion and art as well as certain combinations of these subjects are the macro areas of knowledge, pertaining to the study of the natural objects of universe in different ways. Hence, these subjects are qualified as *first order knowledge*. In contrast, philosophy is called the *second order knowledge* because it aims to unify the different first order disciplines. We would start this discussion with a comprehensive account of the following:

- Core functions and objectives of Philosophy
- Key aspects of Ontology (theory of reality)
- Introduction of Theory of Knowledge (epistemology)

Here we will focus on western philosophy only. In the early stage, Greek philosophers proposed the opposite theories about *mind* and *matter* as the fundamental stuff or reality of universe. Subsequently, various conceptions about mind and matter emerged, which can be analyzed using the notion of **worldview,** which is defined here in an original manner.

Thus there are the main worldviews called organic worldview, spiritual worldview, mechanistic worldview and physical process worldview. The second and third both have rational and empirical parts; hence in total there are six worldviews adopted in the history of western thought. It will logically lead to seven theories of reality, such as Idealism (Theism), Deism, Materialism, Intelligent design argument, Pantheism, Empirical Mysticism and Physical Process Reality. These doctrines are introduced here but to be described later in detail. Then the six worldviews would serve as the framework for giving the *basic classification* of the entire range of diverse philosophical doctrines. However, we note that the different philosophical systems of eastern world -- mainly of India and China -- cannot be neatly brought into this classification scheme.

Subsequently, a detailed analysis of the main areas of western philosophy is given as below.

- Idealism (mainly the doctrines of Socrates, Plato, Aristotle, Augustine and Thomas Aquinas)
- Spiritual process philosophy such as pantheism and historicism (Heraclitus, Zeno, Plotinus, Spinoza, Leibniz, Hegel, Teilhard Chardin and Whitehead)
- Deism and Rationalism (Descartes and Immanuel Kant)
- Materialism and Naturalism (John Locke, George Berkeley and David Hume)
- Mysticism, Phenomenology and Existentialism as well as the Physical Process Doctrines like analytic philosophy and linguistic analysis.

This conflict between the prominent doctrines of western philosophy must be resolved. Our deliberations for that purpose would pave the way for the introduction of System Philosophy in my series of books.

World and Reality

So far our focus was on the area of epistemology, which is one of the two principal branches of philosophy. Now it is the right time to consider the other branch namely *ontology* or *theory of reality* that deals with the question: what is the reality or fundamental stuff of the universe? We can realize at the outset that the eclectic term *reality* gets only very scanty mention in the books and dictionaries on philosophy as well as science and religion; such omission is due to the controversies and conflicts between the various conceptions about reality. Of course, we commonly use the word *reality* in the most abused and confused manner during various situations. Most of us talk about different kinds of realities, rather than *the* reality singularly.

Here it is important to clarify the meaning of *reality* by contrasting it with term *phenomenon*. We can define phenomenon as any object that depends on another object through cause-effect relationship. On the other hand, **reality is the original cause of all phenomena taken as a whole.** Accordingly, reality is self-caused, infinite and permanent. The terms like *ultimate reality* and *ultimate truth* are commonly used as synonyms of reality.

Based on the above points, we want to develop *ontology (theory of reality)* by asking the four questions:

- ❖ Can we conceive reality as a being (substance) or process having existence?
- ❖ What is the method of knowing reality?
- ❖ Specifically, can we know reality from the features of phenomenal objects?
- ❖ How do we link the notion of reality with the religious definition of God?

Then we would analyze the controversies through the insights of System Philosophy pertaining to physical and biological phenomena. We aim to find a new principle of synthesis, which would serve as the summit of our philosophical quest.

In the history of western Philosophy, the exponents of ontology were divided into two groups – first group advocated that consciousness is reality while the second group upheld matter as reality. Hence we may adopt the phrase **monistic philosophy** to denote the erstwhile philosophy as against the newly proposed System Philosophy. Then we establish that, based on *six worldviews,* western monistic philosophy has proposed seven *theories of reality* as mentioned earlier. We must solve the problem of the above ontological divisions. [*].

Our analysis through System philosophy shows that nature as a whole is a *system* composed of three subsystems -- inanimate system, biological system and mental system. It can be proved that these are the successive levels of combinations of matter and consciousness. For defining the structure of ultimate reality we may replace the word *matter* with *body* and treat consciousness in the real sense. Thus we get the definition: *Ultimate Reality is the system of opposite forces called Body and Consciousness, which are represented by X-axis and Y-axis respectively;* it is called the *System Model of Ultimate Reality.* Then reality is like a factory process with opposite components of infinite measure. This model can explain the existence of seven manmade social systems also. [*]

The crucial problem arising here is how to justify the system model of reality. We can show that the proposed *System Model* is a theory, which is verified by clear evidences from universe. Thus it solves the philosophical dilemmas in this regard. [*]

Further, our ethical point of view holds that Ultimate Reality exists as a system with good and bad parts; this expediently solves the *problem of evil,* which troubled thinkers throughout the history of philosophy. The system model has profound implication for the issue of existence of God and Evil. This aspect has already been mentioned in the System Philosophy of God and Evil included in section 7.1.

In this context, it is important to realize that the system model can be interpreted according to process view also. We can modify the central idea of pantheism – mainly neo-Platonism and philosophy of Spinoza – by interpreting that the point of origin pertaining to our system model represents the pantheist notion of reality. Similarly, the logical problem of Brahman-Maya doctrine, under the Vedanta philosophy of Hinduism, is also solved using our system model; it illustrates Maya as the X-axis while Brahman is the Y-axis, implying complementary existence. [*]

Existence of Seven Life Systems

We have already deliberated about the existence of mind, which stands for our mental world including diverse kinds of experience and knowledge. Now we move on to study the human actions that constitute social systems. *We want to shift our philosophical analysis from natural world to social world.* This new focus is vitally important in the context of enquiring about the good and bad aspects of our voluntary actions as well as behavior. [# 5]

The System Philosophy of Ultimate Reality explains the existence of things in natural world, which is technically divided into inanimate world, biological world and mental world. Further, the knowledge about natural things comes from the perspective of an individual human being who exists at the highest level of biological world. In this view, the *mental world* consists of the feelings, ideas and memories happening in the mind of a person; it includes the various emotions like love, hatred, kindness, courage, competition and cooperation also. We may extent the scope of mental world by adding the lower level of mental states occurring in animals, birds and other organisms also. However, as per the practice, we propose to keep human beings at the center of our present philosophical work.

In most of the cases, the mental events occurring in the mind of an individual person would lead to certain *behavior or action*, which can be classified into three kinds - biological, personal and social.

We know that there are certain patterns in the social behaviors of people, though it can vary according to different periods and situations.

Also human beings have established many *social institutions* like family, economy, government and religion. These institutions govern the social behaviors of people at a point of time. Accordingly, different levels of human groups are formed, which together constitute the **social world**. It is pertinent to say that social world as a whole is a man-made entity. And, as far as human life is concerned, social world is the fourth level of universe, which exists above natural world consisting of the levels of inanimate world, biological world and mental world. The scientific knowledge about social world is generally called as **social science**. But can we use scientific method to study the social behaviors in the fields of religion and art?

We devote present chapter to study the philosophical aspects – epistemology and ontology – of the knowledge about various aspects of social world. The central part of our enquiry is concerned with the origin and existence of social institutions as per the layered view of universe. Here we deal with the philosophical aspects of the knowledge about *social world* consisting of different levels of human groups formed by social relations. There are three sections as following:

- Social world and its components
- Epistemology of Social Science
- Seven Life Systems and Knowledge

The scientific study of social world is the area of Social Science. Specifically, sociology is a branch developed on the basis of the main concepts such as group, norms, culture, role or social status, community, social organization, society and institution. If we adopt the methodology of *logical positivism,* it leads to the conflict between rationalism and empiricism. Similarly, we consider the outstanding issues of justification also. [*]

There are numerous social institutions which preexist for imposing the values and goals to social relationships. Tracing the origin of institutions to the dawn of human civilization, we can introduce the innovative idea of *seven life systems* to refer to non-overlapping institutions at the global level. The names are proposed as below:

1. Natural Life System (NLS),
2. Economic Life System (ELS),
3. Political Life System (PLS),
4. Family Life System (FLS)
5. Ethical Life System (ETLS),
6. Artistic Life System (ALS)
7. Religious Life System (RLS)

Each of these *life systems* exists by the dualistic relation between self-interest (SEI) and society-interest (SOI) pertaining to the concerned faculty of human mind. This principle will lead us to the *system model of knowledge* with regard to separate social systems. Accordingly, our deliberation will lead to the epistemological synthesis of the entire spectrum of human knowledge. [*]

Comprehensive View about Truth

The present aim is to solve the dilemmas associated with the notion of truth. We note that truth is an additional condition of a justified belief for becoming knowledge. So we can adopt the definition: **Truth is the property of a justified belief that it corresponds to an actual *state of affairs of the universe*.** It may be recalled that the phrase *state of affairs* denotes a thing or event actually existing in this universe, irrespective of whether any particular person is aware of it. Accordingly, truth is the relation between a proposition and an object in the universe. Here lies the philosophical dilemma.

Our knowledge (proposition) about a thing is solely based on our sense experiences occurring in our mind. It is expedient to treat mind as similar to the screen in a movie theater; images about external object appear in mind just like the characters playing their roles on the screen. The knowing self is analogous to the spectator of the happenings on the screen. Hence mind functions as a screen placed between self and external object. We cannot go beyond the screen of our mental states including images, feelings and ideas. Since we are unable to access the external object, how can we judge that our propositions correspond to the actual *state of affairs* of the universe? This issue makes truth as the most difficult topic of philosophy requiring great intellectual attention.

The controversy about truth arises because deductive propositions are said to possess *necessary truth* and, on the other hand, inductive propositions have only *contingent truth* because it depends on the validity of evidences collected so far. Further, the nonscientific propositions of religion and art call for alternative criterion of truth.

Deductive propositions are normally classified into Theory, Hypothesis and Deduction. These are connected by the logical method of syllogism that is infallible; hence deductive propositions are said to possess *necessary truth*. On the other hand, inductive propositions pertain to Testing and Induction, where the evidences cannot be absolutely certain. Such propositions can have only fallible justification, because it suffers from the problem of induction. In this situation, the truth of inductive propositions is qualified as *contingent truth*, because it depends on the validity of evidences collected so far. Further, in any discipline, there are different shades of necessary truth and contingent truth, in accordance with various *worldviews* adopted. Thus, in the case of science for example, we can differentiate between the propositions under mechanistic worldview and physical process worldview.

Similar study of truth is possible with respect to religion and other subjects. More specifically, in contrast to scientific truth, it is necessary to consider the truth applicable to propositions in religion also. Since the meaning of religious words goes beyond the literal and ordinary sense, we need a higher theory of truth. *More importantly, we desire to find a method for unifying the truths of various branches of knowledge.* In this situation, we have to discuss about the following topics:

- ❖ The Dilemmas about Scientific Truth
- ❖ Issues about Religious Truth
- ❖ System Philosophy of Truth

Through meticulous analysis in the light of system philosophy, we can hold that every kind of knowledge -- that is a TyHDTI scheme -- has ***necessary-contingent truth,*** which is represented by the X-Y model. This innovative theory becomes the *system model of truth* and it reconciles between *necessary truth* and *contingent truth* in the propositions. Moreover, we can now show that scientific truth and religious truth are two levels of truth, just like two technologies of a factory.

In the course of this study, it is necessary to examine the popular view that *truth* is intrinsically related to the notion of *good,* which is central to ethics. Can we deduce that truth and goodness are essentially the same? The question like what is good and what is bad would enlighten us to reach the definition of truth in tune with the comprehensive vision of reality. In this pursuit, the tenets of System philosophy are expected to help us in evolving a unified definition of truth by linking the various kinds of factual truth to the property of value.

This central problem of ethics is solved using the system model. Effectively, we are able to arrive at the definition of *good* and *bad* in the case of propositions, on the basis of the dual purposes – SEI and SOI – of the person generating knowledge. This original treatise has two main advantages. Firstly, it is the method for linking fact with value. Secondly, the ontological existence of truth and falsehood can be illustrated by the *system model of truth*.

Glimpse of the Path Ahead

The field of social sciences as well as natural sciences is inflicted with serious issues of ethics. How can we link fact with value in the case of such study? The traditional philosophy (including analytic philosophy), which prevailed in the hitherto history of thought, has tragically failed in this respect. However, the *system model of truth* enlightens us. Another achievement is the theory of justification of social knowledge on the basis of the principle of *seven life systems*. These ground breaking ideas would prepare us to take the project of System Philosophy to greater heights. The further path is the formulation of integrative philosophy for specific social sciences such as economics, politics and ethics. The comprehensive development of these topics is reserved for a later occasion.

NOTES of Chapter 7

#1 The main references used here are: Alan Grafen and Mark Ridley (2007), Ammar Al Chalabi, et al (2007), Behe (1996), Beird (2003), Brennan (2005), Capra (1983), Capra (1997), Chardin (1965), Darwin (1859), Dawkins (1976), Dawkins (2007), Dawkins (2009), Dennett (1991), Mayr (1999), Griffin (2000), Guttman (2007), Guttman et al (2006), Haught (2000), Haught (2001), Heil (2003), James et al (1987), Jantsch (1989), Job Kozhamthadam (editor) (2004), Kant (2003), Leahey (2005), Lewens (2007), Miller (Editor) (2001), O'Leary (2004), Robert John Russell (Editor) (2004), Shaffer (1994), Tarnas (1991), Thomas (General Editor) (2012), Vijayakumaran Nair and Jayaprakash (2007).

#2 The special references used for discussing the issues under philosophy of mind are: Brennan (2005); Feser (2006); Grayling (Editor) (1995); Guttman (2007); Heil (2003); Leahey (2005); Maslin (2001); Solso (2005)

#3 This author is especially indebted to the following books of reference:

Anthony Harrison– Barbet (1990), Armstrong, Karen (1998), Caputo (2013), Copleston, Frederick S.J. (1994), Davies, Brian (2000), Dawkins (2007), Esposito, et al (2008), Frost. S. E (1989), George Thomas White Patrick (1978), Grayling A.C. (Editor) (1995), Grayling A.C. (Editor) (1998), Griffin (2000), Hick (1994), Kant (2003), Lavin.T. Z. (1989), Macquarrie

(1985), Masih.Y (1995), Max Charlesworth (2006), Tarnas (1991), Thilly (2000), Thomson (1997).

#4 After 1950, there is a proliferation of writings about science-religion problem under the views of both theism and process theology. The main reference books regarding this field are Griffin (2000), Haught (2000), Haught (2001) and Robert John Russell (Editor) (2004). However, the synthesis of science and religion will be achieved in this section by the total integration of traditional philosophies of idealism, materialism and process thought, without resorting to realism; this is my original idea.

#5 The main reference books used for this chapter are the following:

Alex Inkeles (1993); Francis Abraham (1993); Hicks, John (1979); Levin, William C (1984); Lipsey, Richard (1983); Lucy Mair (1992); Martin Hollis (2000); Michael Haralambos and Robin Heald (1990).

Bibliography

Alan Grafen and Mark Ridley (2007). *Richard Dawkins, How a Scientist Changed the Way We Think* (Oxford University Press, Paperback Edition)

Alex Inkeles (1993), *What is Sociology? – An Introduction to the Discipline and Profession*, (Prentice Hall of India Limited, New Delhi, Tenth Indian Reprint, 1993)

Ammar Al Chalabi, Martin R. Turner and R. Shane Delamont (2007), *The Brain – A Beginner's Guide*, (One World- Oxford, England, First South Asian Edition, 2007)

Anthony Harrison– Barbet (1990), *Mastering Philosophy*, (Macmillan, London, 1990).

Armstrong, Karen (1998), *A History of God* (Arrow Books, London, 1998)

Augustine Perumalil (2000), *Critical issues in the Philosophy of Science and Religion* (Indian Institute of Science and Religion, Pune and ISPCK, Delhi, 2006).

Behe, Michael J. (1996), *Darwin's Black Box: The Biochemical Challenge to Evolution*, (New York: Touchstone Books, 1996).

Bird, Alexander (2003), *Philosophy of Science*, (Routledge, London, Indian Reprint, 2003).

Beiser, Arthur (2002), *Concepts of Modern Physics*, (Tata McGraw-Hill, New Delhi, Sixth Edition, Second Reprint, 2002)

Blackburn, Simon (1996), *The Oxford Dictionary of Philosophy* (Oxford University Press, 1996)

Brennan, James F. (2005), *History and Systems of Psychology*, (Pearson education, Delhi, first Indian reprint, 2005)

Brooke Noel Moore and Kenneth Bruder(2005), *Philosophy - The Power of Ideas* (Tata McGraw-Hill Publishing Co. Ltd, New Delhi, sixth edition, 2005.)

Capra, Fritjof (1983), *The Turning Point* (Flamingo, London, 1983).

Capra, Fritjof (1992), *The Tao of physics* (Flamingo, London, Third edition, 1992)

Capra, Fritjof (1997), *The Web of Life,* (Flamingo, London, 1997).

Caputo, John D (2013), *Truth – Philosophy in Transit* (Penguin books, 2013)

Chardin, Teilhard de (1965), *The Phenomenon of Man,* (Harper Torchbook Edition, New York, 1965)

Chatterjee, Margaret (1988), *Philosophical Enquiries,* (Motilal Banarsidas, Delhi, 1988)

Copleston, Frederick S.J. (1994), *A History of Philosophy. Vol. I – IX* (Image Books, Doubleday, 1994).

Darwin, Charles (1859), *On The Origin of Species* (Dover Edition, New York, 2006)

Davies, Brian (2000), *Philosophy of Religion* (Oxford University Press, contains the reprint of Hume's article 'Of Miracles', 2000)

Davies, Paul (1995), *About Time – Einstein's Unfinished Revolution,* (Penguin Books, 1995)

Davies, Paul (2007), *Cosmic Jackpot – Why Our Universe is Just Right for Life* (Houghton Mifflin Company New York, 2007).

Dawkins, Richard (1976), *The Selfish Gene* (Oxford University Press, Oxford and New York, 1976)

Dawkins, Richard (2007), *The God Delusion* (Black Swan, Transworld Publishers. London, 2007)

Dawkins, Richard (2009), *The Greatest Show On Earth: Evidence for Evolution* (Bantam Press, 2009)

Dennett, Daniel (1991), *Consciousness Explained* (Boston: Little, Brown, 1991).

Esposito, John, et al (2008), *Religion & Globalization: World Religions in Historical Perspective.* (Oxford University Press, New York, 2008)

Ewing A. C. (1994), *The Fundamental Questions of Philosophy,* (Allied Publishers Limited, New Delhi, 1994)

Feser, Edward (2006), *Philosophy of Mind* (Oneworld, Oxford, 2006).

Francis Abraham (1993), *Modern Sociological Theory* (Oxford University Press, New Delhi, Ninth Impression, 1993).

Frost. S. E (1989), *Basic Teachings of the Great Philosophers,* (Anchor Books, Doubleday, New York, 1942/1989)

George Thomas White Patrick (1978), *Introduction to Philosophy* (Surjeet publications, Delhi, 1978).

Grayling A.C. (Editor) (1995), *Philosophy: A Guide Through The Subject*, (Oxford University Press, London, 1995).

Grayling A.C. (Editor) (1998), *Philosophy 2 : Further Through The Subject*, (Oxford University Press, London, 1998).

Green, Brian (2005), *The Fabric of the Cosmos*, (Wintage Books, New York, 2005)

Gribbin, John (2008), *The Universe: A biography*, (Penguin Books, London, 2008)

Griffin, David Ray (2000), *Religion and Scientific Naturalism* (State University of New York, 2000)

Grolier Encyclopedia of Knowledge, (Grolier Incorporated, Danbury, Connecticut, 1993). Volumes 1 – 20.

Guttman, Burton (2007), *Evolution– A Beginner's Guide*, (One World- Oxford, England, First South Asian Edition, 2007)

Guttman, Griffiths, Suzuki and Cullis(2006), *Genetics– A Beginner's Guide*, (One World- Oxford, England, First South Asian Edition, 2006)

Guyer, Paul (2008), *Kant*, (Routledge, London and New York, 2006; first Indian Reprint, 2008)

Haught, John F. (2000), *God After Darwin.-A.Theology of Evolution*, (Westview Press, USA, 2000)

Haught, John F. (2001), *Responses to 101 Questions on God and Evolution*, (Paulist Press, USA, 2001)

Hawking, Stephen W. (1995), *A Brief History of Time*, (Bantam Books, 1995 edition)

Hawking, Stephen W. (2011), *The Grand Design*, (Bantam Books, 2011 edition)

Heil, John (2003), *Philosophy of Mind* (Routledge, London and New York; Indian reprint, 2003),

Hick, John H. (1994), *Philosophy of Religion* (Prentice Hall of India Pvt. Ltd, fourth edition, 1994).

Hicks, John (1979), Causality in Economics, (Basil Blackwell Oxford, Great Briton, 1979)

Hospers, John(1997), *An Introduction to Philosophical Analysis* (Allied Publishers Limited, Mumbai, 1997; Original publication by Prentice–Hall in 1953)

Ivor Leclerc (1958), *Whitehead's Metaphysics* (George Allen and Unwin Ltd, London, 1958).

James.T.Shipman, Jerry.L.Adams and Jerry.D.Wilson (1987), *An Introduction to Physical Science* (D.C. Heath and company 1987).

Jantsch, Erich (1989), *The Self-organizing Universe*, (Pergamon Press, 1989).

Job Kozhamthadam (editor) (2002), *ContemporaryScience and Religion in Dialogue- Challenges and Opportunities* (ASSR Publications, Jnana-Deepa Vidyapeeth, Pune, 2002),

Job Kozhamthadam (editor) (2003), *Science, Technology and Values* (ASSR Publications, Jnana-Deepa Vidyapeeth, Pune, 2003),

Job Kozhamthadam (editor) (2004), *Religious Phenomena in a World of Science* (ASSR Publications, Jnana-Deepa Vidyapeeth, Pune, 2004),

Job Kozhamthadam (editor) (2005), *ModernScience, Religion and The Quest for Unity* (ASSR Publications, Jnana-Deepa Vidyapeeth, Pune, 2005),

Job Kozhamthadam (editor- in- chief), *Omega– Indian Journal of Science and Religion* (Institute of Science and Religion, Aluva,)

Kant, Immanuel (2003), *Critique of Pure Reason*, translated by J. M.D. Meiklejohn (Dover Publications, New York, 2003)

Kuhn, Thomas (1970), *The Structure of Scientific Revolutions* (University of Chicago Press, 1970)

Kukla, Andre (1998), Studies *in Scientific Realism*, (Oxford University Press, 1998)

Ladyman, James (2002), Understanding Philosophy of Science (Routledge, London, 2002)

Lavin.T. Z. (1989), From Socrates to Sartre (Bantam Books, New York, 1989)

Leahey, Thomas Hardy (2005), *A History of Psychology – Main currents in Psychological Thought* (Pearson education, Delhi, first Indian reprint, 2005)

Levin, William C (1984), *Sociological Ideas,* (Wadsworth Publishing Company, California, 1984)

Lewens, Tim (2007), *Darwin* (Routledge, London and New York, 2007)

Lipsey, Richard (1983), An Introduction to Positive Economics, (ELBS Edition, 1983).

Lucy Mair (1992). *An Introduction to Social Anthropology* (Oxford University Press, New Delhi, Seventh Impression,1992)

Luke George (2004), *Saptaloka Darshanam- Samgraham*, (PGL Books, Changanachery, Kerala, 2004, in Malayalam language).

Luke, George (2015), *Jeevanum Parinamavum*, (PGL Books, Changanachery, Kerala, 2015, in Malayalam language).

Macquarrie, John (1985), *In Search of Deity – An Essay in Dialectical Theism*,(Cross road Publishing, New York, 1985)

Martin Curd and J. A. Cover (1998), *Philosophy of Science: The Central Issues*, (W. W. Norton & Company, New York / London,1998)

Martin Hollis (2000), *The Philosophy of Social Science* (Cambridge University Press, First Indian Paperback Edition, 2000)

Masih .Y. (1995), *Introduction to Religious Philosophy* (Motilal Banarsidas, Delhi, reprint 1995)

Maslin K.T. (2001), *An Introduction to the philosophy of Mind* (polity Press, UK & USA, 2001)

Max Charlesworth (2006), *Philosophy and Religion- From Plato to Postmodernism* (Oneworld, Oxford, First South Asian Edition, 2006)

Mayr, Ernst (1999), *This Is Biology*, (Universities Press India Limited, Hyderabad, Indian edition 1999)

McGinn, Colin (1998), *The Character of Mind* (Oxford University Press, 1998

Michael Haralambos and Robin Heald (1990), *Sociology - Themes and Perspectives*, (Oxford University Press, New Delhi, Tenth Impression,1990)

Michio Kaku and Jennifer Thompson (2007), *Beyond Einstein: The Cosmic Quest for the Theory of the Universe*, (Oxford University Press, New Delhi, 2007)

Miller, James B. (Editor) (2001), *An Evolving Dialogue-Theological and scientific Perspectives on Evolution* (Trinity Press International, Harrisburg, Pennsylvania, 2001)

Newton, Roger (2010), *The Truth of Science – Physical Theories and Reality* (Viva books, New Delhi, reprint 2010)

O'Leary, Denyse (2004), *By Design or by chance?* (Augsburg Books, Minneapolis, 2004).

Panda N. C. (1999), *Maya in Physics* (Motilal Banarsidas, Delhi, reprint 1999)

Parthasarathy A. (2000), *Vedanta Treatise* (Vedanta Life Institute, Mumbai, 2000)

Robert John Russell (Editor) (2004), *Fifty years in Science and Religion – Ian G. Barbour and his Legacy*, (Ashgate Publishing Ltd, England and USA, 2004),

Robert John Russell, Nancy Murphy and C. J. Isham (Editors) (1999), *Quantum Cosmology and the Laws of Nature : Scientific Perspectives of Divine Action* (Vatican Observatory Publications, Vatican City State and The Center for Theology and the Natural Sciences, Berkeley, California; second edition, 1999)

Romijn, Herms (2002), *Are Virtual Photons the Elementary Carriers of Consciousness?* (Journal of Consciousness Studies, 9, No.1, 2002, pp 61-81).

Rosenberg, Alex (2000), *Philosophy of Science, a Contemporary Introduction* (Routledge, London and New York, 2000),

Russell, Bertrand (1992), *The Problems of Philosophy*, (Oxford University Press, 1992).

Sarojini Henry (2009), *Science Meets Faith* (St. Pauls, Mumbai 2009)

Schilpp, Paul Arthur (Editor) (1941), *The Philosophy of Alfred North Whitehead* (North Western University, 1941)

Schmidt, Paul F. (1967) *Perception and Cosmology in Whitehead's Philosophy*, (Rutgers University Press, New Jersey, 1967)

Sen, Amartya (1990), *On Ethics and Economics*, (Oxford University Press, Delhi, 1990)

Shaffer, Jerome A. (1994), *Philosophy of Mind* (Prentice Hall of India, New Delhi, 1994)\

Sheldrake, Rupert (2013), *The Science Delusion,* (Coronet, Hodder & Stoughton Ltd, London, 2013)

Sherburne, Donald W. (editor) (1965), *A Key to Whitehead's Process and Reality* (Indiana University Press, London, 1965)

Smolin, Lee (1998), *The Life of Cosmos* (Phoenix paperback edition, Great Britain,1998)

Smolin, Lee (2008), *The Trouble with Physics*, (Penguin Books, London, 2008)

Solso, Robert L. (2005) *Cognitive Psychology* (Pearson education, Delhi, second Indian reprint, 2005)

Sweet, William (2003), *Religion, Science and Non-science*, (Dharmaram Publications, Bangalore, 2003)

Tarnas, Richard (1991), *The Passion of the Western Mind* (Pimlieo, London, 1991),

Taylor, Richard (1994), *Metaphysics*, (Prentice- Hall of India Pvt.Ltd, New Delhi, 1994)

Thilly, Frank (2000), *A History of Philosophy*, (SWP Publishers, New Delhi, 2000).

Thomas.A. P, (General Editor) (2012), *Cell and Molecular Biology – The Fundamentals* (Green Leaf Publications, Kottayam, Kerala, 2012)

Thomson, Mel (1997), *Philosophy of Religion* (Teach Yourself Books, UK, 1997)

Urmson J.O. and Jonathan Ree (1989), *The Concise Encyclopedia of Western Philosophy and Philosophers* (Unwin Hyman, London, 1989).

Vijayakumaran Nair and Jayaprakash (2007), *Cell Biology Genetics Molecular Biology* (Academica, Thiruvananthapuram, Fourth Edition)

Washburn, Phil (1997), *Philosophical Dilemmas: Building a Worldview,* (Oxford University Press, New York, 1997)

Whitehead, Alfred North (1978), *Process and Reality* (Original Edition 1929; Corrected Edition by David Ray Griffin and Donald Sherburne, New York, The Free Press,1978)

Index of Names

Alan Grafen, 202

Alex Inkeles, 203

Ammar Al Chalabi, 202

Anthony Harrison– Barbet, 146, 202

Aquinas, Thomas, (1224-1274), 194

Aristotle (BC 384-322), 66, 115-7, 135-7, 145, 154, 168, 182, 194

Armstrong, Karen, 202

Augustine, St. (354-430), 194

Augustine Perumalil, 147

Ayer, A.J. (1910-1989), 93

Behe, Michael, 202

Beiser, Arthur, 27, 29

Bell, John, 28

Berkeley, George (1685-1753), 72, 102, 103, 109, 194

Bertrand Russell (1872-1970), 109

Bird, Alexander, 27, 28, 108

Bohm, David, 17

Bohr, Niels, 11, 17, 28, 122,

Brennan, James F., 202

Broglie, Louis de, 11

Capra, Fritjof, 27, 29, 54, 73, 162-4, 181, 202

Caputo, John D, 202

Chadwick, James, 10

Chardin, Teilhard de, 194, 202

Comte, Auguste (1798-1857), 91

Copernicus (1473-1543), 8, 138

Copleston, Frederick S.J., 202

Cover J A, 27

Darwin, Charles (1809-1882), 27, 178, 186, 202

Davies, Brian, 202

Davies, Paul, 27, 39, 40, 42, 60, 72-4, 162, 166, 180-2

Dawkins, Richard, 186, 202

Democritus, 9

Dennett, Daniel, 202

Descartes (1596-1650), 8, 9, 29, 87, 108, 120, 182, 194

Duhem, Pierre, 109

Einstein, Albert (1879-1955), 14-17, 24, 28, 33, 35, 50, 56, 62, 63, 66, 122, 130, 152, 155-8, 164, 166-7, 180

Esposito, John, 202

Feser, Edward, 202

Feyerabend, Paul, 145

Francis Abraham, 203

Friedman, 32

Frost. S. E, 202

Gamow, George, 33

Galen (AD second century), 138

Galileo (1564-1642), 8, 9, 150

George Thomas W. Patrick, 202

Georgi, 22

Glashow, 22

Grayling A.C, 108, 144, 146, 202

Green, Brian, 27, 39, 40, 42, 71-5, 146, 157, 162, 180-2

Gribbin, John, 39, 72-5, 100, 109, 162

Griffin, David Ray, 202, 203

Guttman, Burton, 202

Guth, Alan, 37, 54

Hartle, J. B., 40

Haught, John F, 202

Hawking, Stephen W (1942-2018), 27, 39, 40-2, 60, 73-4, 129, 146, 157, 162, 165-6, 179, 181-2

Hegel, G. Wilhelm (1770-1831), 194

Heisenberg, Werner, 16, 17, 28, 122

Hempel, Carl (1905-1997), 85-6, 93, 117,118

Heil, John, 202

Hick, John H, 202, 203

Hicks, John, 203

Hippocrates (5th century B. C.), 138

Hospers, John, 108

Hoyle, Fred, 40

Hubble, Edwin, 32-4, 66, 133

Hume, David (1711-1760), 88-9, 103, 109, 118-9, 121, 126-8, 146, 194

James.T.Shipman, 202

Jantsch, Erich, 202

Jayaprakash, 202

Job Kozhamthadam, 72, 147, 181, 202

Jonathan Ree, 108, 109, 145

Kant, Immanuel (1724-1804), 90, 108, 120, 128, 134, 146, 153, 163, 180, 186, 194, 202

Kuhn, Thomas (1922-1996), 123-5, 145

Lakatos, Imre, 145

Laszlo, Ervin, (*see* Preface of this book)

Lavin.T. Z., 202

Leahey, Thomas Hardy, 202

Lederman, Leon M., 55, 73, 74

Leibniz, Gottfried Wilhelm (1646-1716), 194

Lemaitre, 32

Leucippus, 9

Levin, William C, 202

Lewens, Tim, 202

Linde, Andre, 40

Lipsey, Richard, 203

Locke, John (1632-1704), 194

Lucy Mair, 203

Luke, George (1953 -), (*see* Prologue of this book)

Mach, Ernst (1838-1916), 91

Macquarrie, John, 202

Mark Ridley, 202

Martin Hollis, 203

Masih .Y., 203

Maslin K.T., 202

Max Charlesworth, 203

Maxwell, James Clerk (1831-79), 15

Mayr, Ernst, 202

Michael Haralambos, 203

Mill, J. S., 109

Miller, James B., 202

Nambu, Yoichiro, 50

Newton, Isaac (1642-1727), 9, 10, 17, 28, 68, 113, 150, 156

Newton, Roger, 27, 145

O'Leary, Denyse, 202

OMEGA, 72, 181

Oommen. M .A., (*see* Prologue of this book)

Panda N. C., 182

Pauli, 13

Penrose, Roger, 40

Penzias, Arno, 33

Planck, Max, 14, 35

Plato (BC 428-348), 145, 194, 197

Plotinus (AD 205-270), 194

Podolsky, 28

Popper, Karl (1902-1994), 95-7, 145

Ptolemy (about AD 150), 8, 137, 138

Quine, W.V.O., 109, 145

Robert John Russell, 202, 203

Robin Heald, 203

Rosen, 28

Rosenberg, 27, 107-9, 145, 146

Rutherford, 10, 11

Salam, 22

Schlick, Mortiz (1882-1936), 93

Schrödinger, 17, 28

Shaffer, Jerome A., 202

Sheldrake, Rupert, 147, (*see* Preface also)

Smolin, Lee, 39, 42, 53, 71, 73, 75

Spinoza, Benedict (1632–1677), 194, 197

Solso, Robert L., 202

Tarnas, Richard, 27, 108, 202, 203

Teresi, Dick, 55, 73, 74

Thilly, Frank, 203

Thomas., 202

Thomson, J. J., 10

Thomson, Mel, 203

Urmson J.O., 108, 109, 145

Vijayakumaran Nair, 202

Virk, Hardev Singh, (*see* Introduction of this book)

Washburn, Phil,

Weinberg, 22

Whitehead, Alfred North (1861-1947), 122, 194

Wilson, Robert, 33

Young, Thomas (1773-1829), 15

Zwicky, Fritz, 64

Index of Subjects

abbreviation, 88-9, 102, 121, 127, 153, 192

accidental regularities, 80-1, 90

algorithm, 18, 104, 170, 185

anthropic principle, 166, 177, 179

antigravity, 50, 53-4, 66-7, 121, 176, 182

antirealism, 122

a priori, 82, 108

allopathy, 138

arrow of time, 132, 151, 180

astronomy, 5, 14, 27, 31, 41, 137

astrophysics, 67, 149, 150, 154, 156-7

atheism, 106, 173, 186, 190

atom / atomic theory, 4, 6, 9-29, 32, 36, 44-5, 50, 62, 78, 119, 120, 126, 154, 168

atomism, 9

Ayurveda, 138-140, 146,

background radiation, 33

balloon model, 32-3, 39, 63, 133, 146

basic forces, definition of, 12-22, 39

Bayesianism, 91

behaviorism, 95, 106, 114, 188

Bell's Theorem, 28, 29

Bible, 27, 165

big bang, 7, 27-42, 130-133, 170-5

biological evolution, 184-186

black hole, 42, 65, 73, 153-165

body-mind (mind-body) dualism, 8, 29, 88-9, 120, 164, 182, 188

Brahman, 163-4, 197

brane, (*see* membrane)

Cartesian dualism, 164

cause-effect relation, 9, 10, 18-9, 77-89, 111, 118, 121, 136-7, 151, 155, 169, 191, 195

CERN, 23, 55

Chinese mysticism, 29

classical science, 6-19, 26, 44, 60, 76, 79-94, 103-122, 139, 151-171

Closed universe, 63-65

CMB, 34, 38

COBE, 33, 34, 38, 62-7

compactification, 59-61, 70, 130, 133, 149, 172-3

complementary principle, 16, 17, 28, 116-7, 125, 166, 178, 189, 197

computer model functionalism, 95, 104, 115, 188

consciousness, 90, 104, 117, 138, 142, 164, 175, 179, 186-196

content view, definition of, 2-4, 94-5, 103, 105-8

contingent truth, 90, 105-6, 117, 144, 200-1

Copenhagen Interpretation, 28

corroboration, 96

cosmic microwave background, 34, 38

cosmological constant, 66, 74, 176-7, 182

cosmological puzzle / question, 27, 39, 42, 58 -67, 111, 148-182

cosmological stages, 44-9, 54, 72, 98, 101, 125, 131, 134

creation, (*see* God)

critical density, 62, 63

dark energy / dark matter, 44, 58, 61- 73, 103, 123-134, 149-181

Darwinism / Darwin's theory, 27, 178, 186, 202

deduction / deductive propositions, 4, 79-121, 137-153, 192, 200

Deductive-Nomological Model (D-N Model), 85, 117

deism, 28, 105-6, 120, 170, 194

diameter of universe, 37, 38, 49, 160

Discovery of Reality, 3, 121, 179, 182-4, 192

doshas, 139

Duhem-Quine Thesis, 109

elementary particles, 21-33, 54-5, 68-70, 98-9, 130, 171

$e = mc^2$, 14, 18, 20, 24, 35, 162-7

empiricism, 86-92, 102-124, 146, 198

Enlightenment, 88

Epiphenomenalism, 89, 90, 104-6

Epistemology, definition of, 25, 27, 76-9, 86

EPR paradox, 28

eternal chaotic inflation, 40

eternal inflation, 60, 74

exchange particle, 20-4, 51, 56-7

existence, (*see* predicate)

existence is an inference, 134, 174

existence of matter, 5, 29, 78, 87, 105-111, 121-128, 149-168

existence of physical world, 131, 134, 153, 166, 176

expansion of universe, 32-39, 48-70, 131-133, 146, 161, 173-6

evolution of universe, 58, 170

falsification theory, 95-7, 145

fine tuning, 166, 177-9

flatness problem, 37

flat universe, 62-65

frequently asked question (FAQ), 148, 170, 179

general theory of relativity, 33, 49-50, 56-66, 156

geocentric theory, 8, 138

geometry of universe, 52, 59, 62-3, 116, 137, 173

God, 4, 9, 10, 17, 27, 35, 42, 55, 61, 93, 106, 113, 115, 126, 136-145, 162-170, 178-191, 195, 196

God particle, 55, 73, 74

good theory, 96, 118, 119

grand unified theory (GUT), 37, 47, 54, 57, 70, 129, 162-164

gravitational force (gravity), 9-20, 42, 44, 47, 49-57

gravitational waves, 148, 156-161

gravitons, 51, 74

group theory, 22, 98, 172

habit of association of ideas, 89, 121

heliocentric theory, 8, 138

Higgs boson, 25, 56-7, 74, 158

Higgs Field, 55-7, 67, 68

Higgs mechanism, 41-57, 67-73, 98, 159-164, 173

holism of meaning, 117, 145

horizon problem, 37

Hubble's discovery, 32-4, 66, 133

hypothesis, 4, 31, 83-102, 112, 192

Hypothetico-Deductive, 85, 117

Idealism, 106, 136-145, 170, 188, 194, 203

Induction / inductive propositions, 4, 29, 81-97, 105-127, 138-147, 197, 200

industrial revolution, 8

inference, definition of, 20, 61, 79

Inference to the Best Explanation (IBE), 125-8, 178

infinite / infinite being, 145, 149, 154, 177-9, 195

inflation / inflationary big bang theory, 30-67, 130, 159-173

instrumentalism, 97, 122-4, 162

intelligent designer, 111, 148-9, 166, 170, 175-182, 194

invisible, 43-4, 55, 63-9, 82-3, 101-3, 116-127, 154-9

justification, 57, 76, 79, 86, 92, 96-7, 104-153, 164, 172-201

Large Hadron Collider (LHC), 23, 55, 98, 107, 160

layered perspective, 154, 180

life, (*see* philosophy of life)

life and mind, 5, 7, 30, 95, 114-6, 170, 172, 178-187

life system, 143-4, 183, 190-201

light quanta, 17, 20

LIGO, 157-161, 173, 181

logical positivism, 76, 86-99, 107-9, 113-122, 198

machine-algorithm model, 18, 185

MACHO, 5, 9-13, 20-32, 49-57

mass, 5, 10-26, 32, 49-57

materialism, 28, 88, 103-6, 114, 121, 125, 140, 145, 162, 170-173, 186, 194, 203

matter and energy, 2-18

matter is not an illusion, 163-169

matter, puzzle of, 71, 166-9

maya, 182, 197

mechanistic worldview, 6-10, 79-93, 112-124, 151-4, 194, 200

membrane, 41, 44, 54, 58-60, 68-74, 98-102, 111, 123, 127, 131-4, 149-165, 172, 176

metaphysical realism, 87, 106, 120, 143, 163-4, 190

methodology, definition of, 8, 31, 76, 79, 80, 85-92

model dependent realism, 165

mophism (*see* modern phenomenalism)

mother-universe, 37

modern phenomenalism, 101-4, 109, 114, 132

M-Theory, 53, 58-61, 130-1

multiverse, 41-4, 54, 58-61, 68-73, 98, 101-2, 131-4, 154-176

mysterious, 44, 58-109, 131, 154

mystery (pagan) religions, 136, 138

mystic mind, 142-4, 178, 182-191

mystical process science, 54, 125, 135, 147, 163-170

mysticism, 3, 17, 29, 54, 101, 106, 170, 190-194

naïve realism, 97, 106, 120-4, 152

natural science, 6, 7, 72, 77-8, 143, 201

necessary-contingent truth, 201

necessary truth, 106, 117, 127, 140, 144, 200-1

negative time, 154, 155

neo-Platonism, 197

nervous system, 89, 104, 117, 139, 187, 188

net energy was zero, 129, 165

neutrino research, 101, 107, 161

no boundary, 37, 40

non-accidental regularity, 81, 84

non-science, 28, 96, 143

nucleosynthesis, 36, 64

observable universe, 33

OMEGA, 72, 181

open universe, 63

opposites, existence of, 116-7, 134-5, 164, 189

origin of universe, 163-175

panchabhutha, 139

pantheism, 29, 106, 136-140, 145, 194, 197

paradigm, 3, 10, 18, 119, 123-4, 145, 155

paranormal phenomena (psi), 140-147

parapsychology, 142-147

particle accelerator, 21, 23

particle-wave duality,15-18, 28, 68-71, 92-101, 113-4, 152-169

past universe, 61, 130-134, 146-155, 173-4

phase transition, 48, 129

phenomenalism, 102

philosophy of life, 185-6

philosophy of mind, 1, 86, 88, 104, 115-7, 143, 182, 187-9

philosophy of religion,1, 145, 178, 189-191, 203

philosophy of science, 1, 10, 17, 28-9, 39, 76-147

physical laws, 7-10, 28, 33, 68, 87-90, 113, 121, 140, 179, 191

physical process worldview, (*see* process view)

physical reality, 17, 43, 68, 97, 101-3, 109, 125, 134, 166, 172

physical science, 1-9, 76, 78, 105-153

Pluralism, 25, 52-3, 101, 123-9, 160, 172

pocket universe, 59-61, 70, 132

positivism, 91, 93

postmodernism, 124

pragmatism, 106, 122

predicate, 120, 128, 134, 146, 153, 180, 186

present universe, 33, 36, 60, 66, 70, 130-4, 146, 149, 173-4, 181

problem of induction, 84, 89, 95-7, 113, 117-127, 178, 200

process view, 2-7, 93-7, 105-135, 163-4, 179, 185, 188-196

progress of science, 82, 105, 123-4

pseudo science, 107, 135, 140-4

puzzle of matter, 71, 166-169

quantum cosmology, 19, 27, 31, 39, 41-75, 92-182

quantum field theory, 1, 2, 14, 19-69, 101, 158, 163

quantum fluctuation, 37, 54, 64, 73

quantum gravity, 41-74, 97-105, 125-134, 159-168, 172

quantum mechanics, (*see* quantum cosmology)

quantum physics, 3, 6, 7, 14-5, 36-7, 95, 122, 125, 153, 171

quantum theory, 6, 14-5, 31, 42

rationalism, 86-91, 103-116, 130, 194, 198

realism, (*see* metaphysical realism / scientific realism)

reality, 2, 3, 29, 54, 78, 80, 105, 112, 120-8, 136, 142-5, 162-4, 170, 175, 179-183, 192-201

reductionism, 10

refined big bang theory (RBBT), 35-40

reformation, 8

reliabilism, 118

religious philosophy, 28

Renaissance, 8, 135

Roman Catholic Church, 35

rules of thought, 115-7, 135, 154, 168-9, 182, 186

sastra, 140-146

science, definition of, 1-29

scientific fiction, 101, 166

scientific method, 34, 39, 76-88, 93-9, 105, 112, 117, 123-4, 142,147, 165, 198

scientific mind, 9, 90, 112, 117, 126-8, 141-4, 172-8, 182, 189, 191

scientific revolution, 8, 123

scientific realism, 106, 109, 123-8, 132, 145, 152, 156, 161-178

scientism, 106-9, 161, 173

singularity, 34-7, 130-3, 149, 174

skepticism, 90-1, 103, 113, 118-129, 161, 172, 178

social institution, 82, 107, 124, 145, 197, 198

social science, 6-8, 28, 78-9, 91, 123, 135, 170, 192, 198, 201

social system, 88, 143, 170, 184, 190, 196-9

social world, 78, 95, 198

soul, 4, 9, 87-8, 93, 103, 115, 138, 178, 187, 189, 191

source, 76-79, 86-9, 95-7, 103-116, 143-5, 149, 191-193

space and time,

- abstract, 154

- curvature of, 156-7, 181

- concept of, 2, 5, 9, 10, 14-6, 21, 33-4, 49, 61, 94, 129, 149-156, 168-175, 180

- quantum, 155

- visible, 154

space-time, 16, 51, 61-66, 102, 130-3, 142, 152-8, 166-172, 181

special theory of relativity, 14. 15, 152, 156

spiritual science, 110, 135-147, 193

standard big bang, 30-7, 40, 49, 69

standard forces, 14, 36, 50-7, 70-5, 129, 130, 163, 165

standard model, 23-9, 42-74, 97-109, 125-130, 159-176

steady state theory, 40

string theory, 42-74, 95-134, 146-172

superstring, definition of, 51-3, 58

strong nuclear force, 31-56

structure of atom, 10, 119, 169

subatomic particle (subatomic phenomenon), 4, 12-29

substance, 10-3, 19-22, 44, 68

supernova, 66, 156

superscientific, 189

syllogism, 83-9, 137, 140, 200

symmetry / symmetry-breaking, 22, 41-57, 70, 72, 98, 129

system, definition of, 116

system model, 110-135, 143-175, 186-190, 196-201

taxonomy of science, 6

teleology, 136

testing, (see inference)

theism, 28, 145, 170, 194, 203

theoretical entities, 43, 82-99, 102, 113-4, 122, 140, 152-3, 172

theory, definition of, 82, 83

TOE, 51, 57, 74-5, 130

theory of four causes, 136

theory of knowledge, (see epistemology)

thought experiment, 28, 152-5

time, (see space and time)

truth, 76-9, 84-97, 104-112, 107, 124, 127, 140, 144, 183-4, 191-201

twin paradox, 155

TyHDTI scheme, definition of, 80-85

Ultimate Reality, 54, 175, 188, 191-7

uncertainty principle, 16, 19, 28, 54, 169

underdetermination, 39, 94-99, 109

unification, 20-2, 42, 47-59, 71, 75, 101, 123, 128, 160, 191

vacuum, 42, 47, 54-5, 67,70, 162-4

Vedanta / Vedantic Cosmology, 163-4, 170, 182, 197

verifiability criterion of meaning / verification principle, 93,

95, 99, 101, 103, 109, 114, 118, 122

virtual particle, (*see* exchange particle)

visible world, 13, 16, 20, 30, 43-5, 67, 92, 103, 116, 122, 126, 151, 154, 168-9

vital force (élan vital), 137

weak nuclear force, 13-4, 17, 20-2, 24, 36

WMAP, 33-4, 38, 62-3, 67

worldview, definition of, 3-9, 18, 28, 79, 82

X-Y model, 117, 131, 167, 173-4, 188, 201

yoga, 138, 141, 146

zero net energy, 129, 165

****

www.ingramcontent.com/pod-product-compliance
Lightning Source LLC
Chambersburg PA
CBHW020741180526
45163CB00001B/301